AF342448

High Reynolds Number Flows
Using Liquid and Gaseous Helium

Russell J. Donnelly

Editor

High Reynolds Number Flows Using Liquid and Gaseous Helium

Discussion of Liquid and Gaseous Helium as Test Fluids

Including papers from The Seventh Oregon Conference on Low Temperature Physics, University of Oregon, October 23–25, 1989

With 180 Figures

Springer-Verlag

New York Berlin Heidelberg London Paris
Tokyo Hong Kong Barcelona Budapest

Russell J. Donnelly
Department of Physics
University of Oregon
Eugene, OR 97403
USA

Library of Congress Cataloging-in-Publication Data
High Reynolds number flows using liquid and gaseous helium :
    discussion of liquid and gaseous helium as test fluids : including
    papers from the Seventh Oregon Conference on Low Temperature
    Physics, University of Oregon, October 23–25, 1989 / sponsored by
    the Defense Advanced Research Projects Agency through the United
    States Office of Naval Research : [edited] by Russell J. Donnelly.
        p.   cm.
    Includes bibliographical references.
    ISBN 0-387-97475-X
    1. Fluid dynamics — Congresses.   2. Reynolds number — Congresses.
3. Helium — Congresses.   4. Liquid helium — Congresses.   I. Donnelly,
Russell J.   II. Oregon Conference on Low Temperature Physics (7th :
1989 : University of Oregon)   III. United States.   Defense Advanced
Research Projects Agency.   IV. United States.   Office of Naval
Research.
QC150.H55   1991
532′.05 — dc20   91-32312

Printed on acid-free paper.

Production managed by Hal Henglein; manufacturing supervised by Jacqui Ashri.
Camera-ready copy prepared by the contributors.
Printed and bound by Edwards Brothers, Inc., Ann Arbor, MI.
Printed in the United States of America.

9 8 7 6 5 4 3 2 1

ISBN 0-387-97475-X Springer-Verlag New York Berlin Heidelberg
ISBN 3-540-97475-X Springer-Verlag Berlin Heidelberg New York

Preface

Liquid helium has been studied for its intrinsic interest through much of the 20th century. In the past decade, much has been learned about heat transfer in liquid helium because of the need to cool superconducting magnets and other devices. The topic of the Seventh Oregon Conference on Low Temperature Physics was an applied one, namely the use of liquid and gaseous helium to generate high Reynolds number flows. The low kinematic viscosity of liquid helium automatically makes high Reynolds numbers accessible and the question addressed in this conference was to explore various possibilities to see what practical devices might be built using liquid or gaseous helium. There are a number of possibilities: construction of a wind tunnel using critical helium gas, free surface testing, low speed flow facilities using helium I and helium II. At the time of the conference, most consideration had been given to the last possibility because it seemed both possible and useful to build a flow facility which could reach unprecedented Reynolds numbers. Such a device could be useful in pure research for studying turbulence, and in applied research for testing models much as is done in a water tunnel.

In order to examine these possibilities in detail, we invited a wide range of experts to Eugene in October 1989 to present papers on their own specialties and to listen to presentations on the liquid helium proposals. While many of our attendees were not known to one another before the conference, it soon became evident that there were exciting things to talk about and that the proceedings of the conference should appear in permanent form. I was fortunate enough to contact Dr. Thomas von Foerster of Springer-Verlag in New York, who agreed to publish the results of the conference as a small volume.

The papers presented here were prepared after the decision to have a published record, and an effort has been made to include ideas which came from the conference and from subsequent research.

As a result the papers here are less of a conference report than contributions to a volume proposing the use of helium as a test fluid. I have written the first article as an attempt to outline what I know on the subject of testing with liquid and gaseous helium, some of which was learned after the conference ended. I have attempted to make this article self-contained as an introduction to the subject. The remaining papers fill out many of the details which I have been unable to cover in my own survey. These papers concern existing facilities and research programs, ranging from testing of models to modern research in turbulence. The final seven papers explore various aspects of low temperature physics which bear on the practical possibility of high Reynolds numbers research with liquid and gaseous helium.

We reproduce an important paper on high Reynolds number wind tunnels by courtesy of Dr. Ronald Smelt of Oakland, Oregon. Smelt wrote this paper in 1945 as Chief of High Speed Flight at the Royal Aircraft Establishment, Farnborough, England. The paper became available only in 1979, 34 years after he wrote it. Now retired from the position of Vice President and Chief Scientist of Lockheed Aircraft Corporation, I came to know him only after this conference took place.

Early support for the idea of generating high Reynolds numbers using liquid helium came from Gary W. Jones of DARPA, who made funds available through ONR in a program administered by Edwin Rood. I am grateful to Gary Jones and Edwin Rood for their support and to Patrick Purtell for continuing support from ONR through grant ONR N00014-89-1274.

The Oregon Conferences on Low Temperature Physics have concentrated on superfluidity, one of the most fundamental subjects in condensed matter physics. It has been a remarkable experience to come to know the engineers whose brilliant work in testing and in turbulence is represented by the papers in this volume.

I hope the discussions presented here will prove to be but the first step in the common goal of engineers and scientists in achieving the highest Reynolds numbers in a controlled environment.

August, 1991   Russell J. Donnelly

Contents

Introduction

Modern Wind Tunnels

Contributors

Russell J. Donnelly

Physics Department
University of Oregon
Willamette Hall
Eugene, OR 97403

Robert Kilgore

Applied Aerodynamics Division
NASA Langley Research Center
MS 285
Hampton, VA 23665-5225

Pierce Lawing

NASA Langley Research Center
MS 267
Hampton, VA 23665-5225

D.M. Bushnell

NASA Langley Research Center
Hampton, VA 23665-5225

G.C. Greene

NASA Langley Research Center
Hampton, VA 23665-5225

Leonard Weinstein

Fluid Mechanics Division
NASA Langley Research Center
MS 163
Hampton, VA 23665-5225

David Coder

David Taylor Naval Research & Dev. Center
Code 1543
Bethesda, MD 20084

S.G. Flechner

NASA Langley Research Center
Hampton, VA 23665-5225

J.B. Peterson, Jr.

NASA Langley Research Center
Hampton, VA 23665-5225

Lisa Bjarke

NASA Ames-Dryden Flight Research Facility
MS D-OFA
P.O. Box 273
Edwards, CA 93523-5000

Michael Goodyer

Department of Aeronautics & Astronautics
University of Southampton
Highfield, Southampton
England

Colin Britcher

Old Dominion University
IMEDD
Norfolk, VA 23529-0247

K. R. Sreenivasan

Department of Mechanical Engineering
P.O. Box 2159 Yale Station
New Haven, CT 06520-2159

Klaus Schwarz

IBM, Thomas J. Watson Research Center
P.O. Box 218
Yorktown Heights, NY 10598

Naoiki Ichikawa

Applied Physics & Information Science
1 - 2 Namiki
Tsukuba-City
Ibaraki 305
Japan

Masahide Murakami

Institute of Engineering Mechanics
Tsubuka University
Tsubuka
Ibaraki 305
Japan

T. Yamazaki

Tsubuka University
Tsubuka
Ibaraki 305
Japan

A. Nakano

Tsubuka University
Tsubuka
Ibaraki 305
Japan

H. Nakai

Tsubuka University
Tsubuka
Ibaraki 305
Japan

Steven Van Sciver

Nuclear Engineering & Engineering Physics
University of Wisconsin-Madison
921 Engineering Research Building
1500 Johnson Drive
Madison, WI 53709-1687

Michael Smith

Department of Physics
University of Oregon
Willamette Hall
Eugene, OR 97403

Joseph Niemela

Department of Physics
University of Oregon
Willamette Hall
Eugene, OR 97403

Glen McIntosh

Cryogenic Technical Services
3445 Penrose Place
Suite 230
Boulder, CO 80301

K.R. Leonard

Cryogenic Technical Services
3445 Penrose Place
Suite 230
Boulder, CO 80301

Ronald Smelt

7250 Driver Valley Road
Oakland OR 97462

Introduction

Liquid and Gaseous Helium as Test Fluids

Russell J. Donnelly
Department of Physics, University of Oregon
Eugene, Oregon 97403

1. Introduction

The purpose of this book is to examine the possibility that liquid and/or gaseous helium at low temperatures could be useful in generating high Reynolds number flows and in testing models. In this paper, I shall take a general look at the different sorts of flow facilities one might consider, and try to identify those which appear promising enough to pursue. The first paper which pointed to the advantage of helium for testing was written by Ronald Smelt (1945). It is reproduced as an Appendix to this volume.

The foundation of testing with models is the idea of dynamical similarity, that is, the examination of the conditions under which two geometrically similar objects have a similar flow pattern. A good introductory discussion of this problem has been given by Tritton (1988). The simplest situation is the incompressible flow of a fluid at velocity U over a body of characteristic dimension L. Here dynamical similarity of two geometrically similar systems exists if they have the same Reynolds number

$$Re \quad = \quad UL/\nu \qquad (1)$$

where ν is the kinematic viscosity. The kinematic viscosity

$$\nu \quad = \quad \eta/\rho \qquad (2)$$

is the ratio of the dynamic viscosity η to the density ρ. Dimensionless parameters such as (1) can be expressed as ratios: for example, the Reynolds number can be interpreted as the ratio of inertial to viscous forces.

The measure of compressibility, the ratio of change in density to density of the fluid at rest is the square of the Mach number

$$Ma \quad = \quad U/c \qquad (3)$$

where c is the speed of sound in the fluid.

Other test situations involve a fluid with a free surface. Here the Reynolds number characterizes the turbulent flow, but other parameters also matter: the ratio of inertial forces to gravitational forces is the square of the Froude number

$$Fr \quad = \quad U/(g\,L)^{1/2} \qquad (4)$$

(where g is the acceleration of gravity) which is important for wave generation. The ratio of inertial forces to surface tension forces is the Weber number

$$We \quad = \quad \rho U^2 L/\sigma \qquad (5)$$

(where σ is the surface tension). Another important parameter is the ratio of the density of the liquid to the density of the gas above it.

For rotating fluids the ratio of inertial forces to Coriolis forces is known as the Rossby number

$$Ro = U/L\Omega \qquad (6)$$

and the ratio of viscous forces to Coriolis forces as the Ekman number

$$Ek = v / \Omega L^2 \qquad (7)$$

The importance of cryogenics in wind tunnels arises primarily because of the need for high Reynolds numbers. When a gas is cooled, the viscosity drops and the density rises: both factors raise the Reynolds numbers that can be achieved.

The largest cryogenic wind tunnel in the world is the U.S. National Transonic Facility (NTF) at the NASA Langley Research Center, Hampton, Virginia. There, Reynolds numbers as high as 10^9 can be achieved by spraying liquid nitrogen into the flow and operating at elevated pressures (~9bar). The NTF is transonic in the sense that the Mach number is high enough (≤ 1.2) that the flow over an airfoil develops a shock wave.

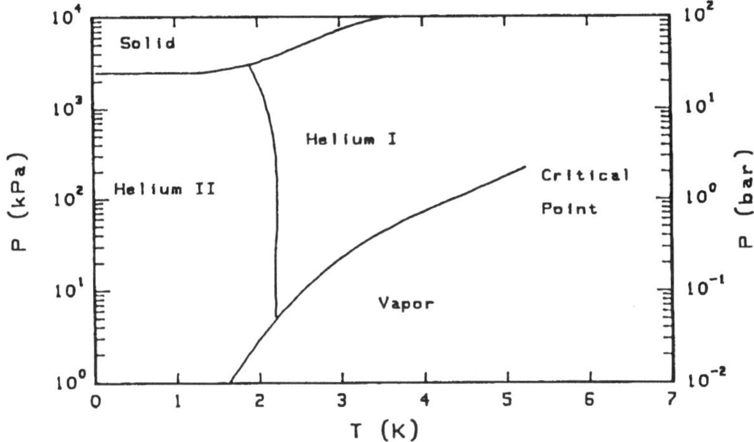

Figure 1 Phase diagram of helium showing the regions of He I, He II and solid helium. The line separating He I and He II is called the lambda line, and the critical point is the neighborhood of conditions for critical helium gas experiment.

The phase diagram of low temperature helium is shown in Figure 1. The critical temperature is 5.2K and the critical pressure is 2.26 bar. The normal boiling point of liquid helium is 4.2K and by means of pumping over the liquid one can reduce the temperature to ~1K. Other means can be found to go lower in temperature, and in fact liquid helium will not solidify under its own vapor pressure at any temperature. It requires about 25 bar to solidify helium at low temperatures.

If we pump on a sample of liquid helium we reduce the pressure, lower the temperature, and follow along the vapor pressure curve in Figure 1. From 5.2 K to 2.172 K we are in a liquid phase called helium I. Experimentally the liquid is seen to boil under reduced pressure just as any normal liquid would. Below $T_\lambda = 2.172$ K, the so-called lambda transition, the liquid enters a special state known as helium II, which exhibits superfluidity. Helium II will be the subject of Section 3 below. Above the lambda transition, the liquid behaves as a classical fluid called helium I: "classical" in the sense that it obeys the Navier-Stokes equation and the expected boundary conditions. Evidence for this will be discussed by Niemela elsewhere in this volume.

The temperature dependence of the kinematic viscosity and surface tension of liquid helium are illustrated in Figures 2 and 3.

Helium gas near the critical point has some extraordinary properties. The viscosity and speed of sound are rapid functions of temperature and pressure as illustrated in Figure 4.

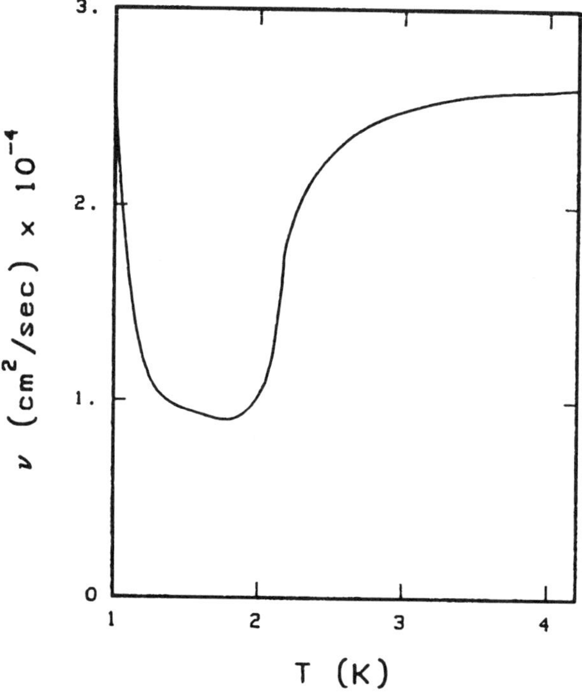

Figure 2. The temperature dependence of the kinematic viscosity of liquid helium.

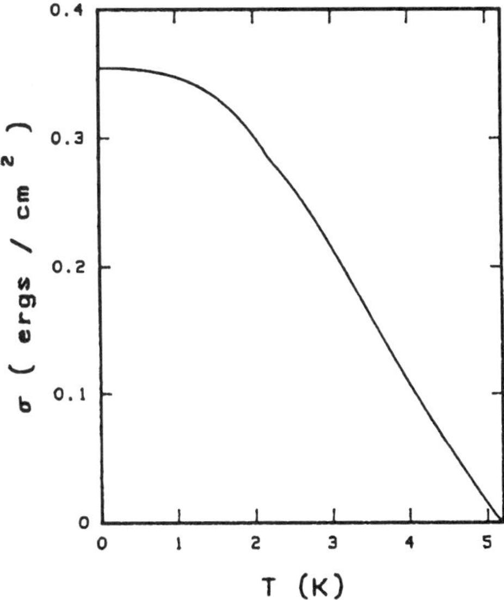

Figure 3. The temperature dependence of the surface tension of liquid helium.

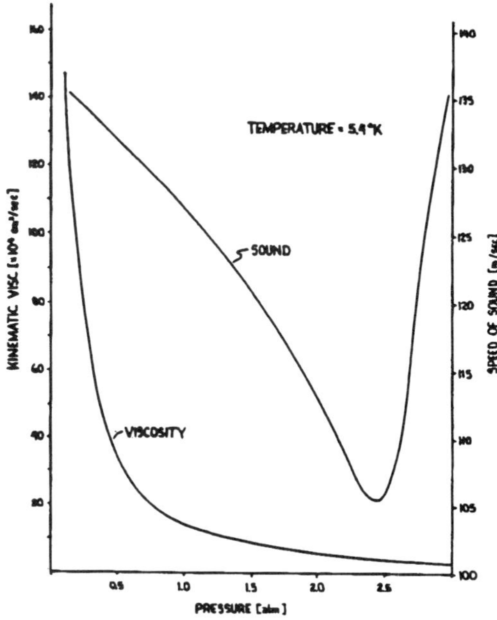

Figure 4. Properties of helium gas at 5.4K as a function of pressure. A similar set of
curves is available for every temperature.

2. Why Helium is a Candidate for High Reynolds Number Research

There are a number of different topics discussed in this article and it is important that the reader distinguish carefully whether the discussion is on helium gas, helium I or helium II.

(a) Critical Helium Gas

The kinematic viscosity of air at $20°C$ is 0.150 cm²/sec. Helium gas, as seen from the data in Figure 4, has a lower kinematic viscosity than air and hence will generate higher Reynolds numbers.

One of the first to realize that helium gas would make a good test fluid was Ronald Smelt, as seen in his paper in the Appendix. Smelt's paper was originally restricted in circulation and was first seen at Langley by Goodyer and Kilgore after their successful use of a low-speed cryogenic nitrogen tunnel in 1972. Articles by Goodyer and Kilgore are incorporated in this volume.

The ratio of the specific heats of air $\gamma = 1.4$. Helium gas, being monatomic has $\gamma = 1.66$ in the ideal limit, changing to other values both greater and less than the ideal limit as the critical point is approached and quantum effects become important.

The first use of helium gas I am aware of in fluid mechanics was, in fact, in the field of thermal convection. Threlfall (1975) and later Libchaber (1987) have been able to cover 11 orders of magnitude in the Rayleigh Number by varying the temperature and pressure of the gas in a container of modest dimensions. We show in Figure 5 data by Libchaber presented as a plot of Nusselt number vs. Rayleigh number. The results of these studies have stimulated great interest in understanding the properties of thermally induced turbulence and are certainly an example of the efficacy of helium gas as a test fluid.

Behringer and Ahlers (1982) have reviewed the heat transport problem in Bénard convection for both critical helium gas and helium I. The results leave no doubt that these are Navier-Stokes fluids.

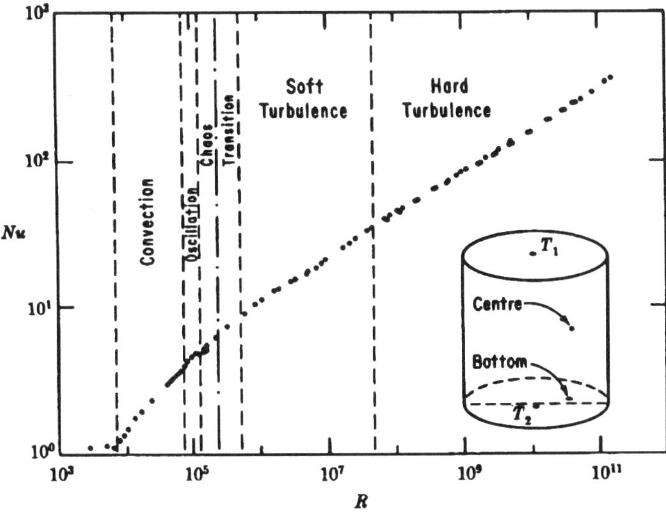

Figure 5 Thermal convection data obtained in critical helium gas by Libchaber (1987). Note that Nusselt number is measured over 11 orders of magnitude in Rayleigh Number by adjustment of temperature and pressure. Latest results have extended even this range.

(b) Liquid Helium

The interest in liquid helium, by contrast to the previous section, is for incompressible flows. The kinematic viscosity of water at $20\,°C$ is $1\text{x}10^{-2}$ cm²/sec and for helium I at 2.18 K is $1.78\text{x}10^{-4}$ cm²/sec. Thus $\nu_{He\,I}/\nu_{H_2O} = 1.7 \times 10^{-2}$ and $\nu_{He\,I}/\nu_{air} = 1.15 \times 10^{-3}$. A flow facility could therefore be 59 times as small for helium I as for water, or 870 times as small as for air at the same Reynolds number. Similar choices could be made for velocity. Since the speed of sound in helium I is about 220 m/sec the Mach numbers for most flows with helium I will be nearly zero. At a flow velocity of 1m/sec over a chord of 20 cm, the Reynolds number will be $\sim 1.16 \times 10^7$ and we can see immediately that a flow facility using helium I as a working fluid will be a low speed (i.e. $\mathrm{Ma} \rightarrow 0$), high Reynolds number device, no different in principle than a water tunnel. (Water tunnels are discussed in this volume by Bjarke.) This single observation is the principal reason for interest in liquid helium I: it should be possible to reach Reynolds numbers for incompressible flows as high as in the NTF with a compact, relatively low cost flow facility.

The kinematic viscosity of helium I and helium II are about the same near T_λ, but helium II exhibits superfluidity and it would be interesting to explore what that means for modelling. After all, ships and aircraft do not move through superfluids.

3. Helium II and the Two Fluid Model. Quantized Vortices

The background on the physics of helium II is contained in two books by the author: Donnelly (1967, 1991). These books contain complete references to authors and their contributions, omitted by necessity in a brief account such as this. When liquid helium viscosity is measured in a rotating cylinder viscometer, the result is of order $20\mu p$ and is temperature dependent. On the other hand helium II will flow through fine capillary tubes with no pressure drop whatever. This apparent *viscosity paradox* is conventionally explained by imagining that helium II consists of some sort of mixture of two fluids: that is, a normal fluid of density ρ_n, velocity v_n and viscosity η, and a superfluid of density ρ_s, velocity v_s and zero viscosity. Thus the total density ρ

$$\rho = \rho_n + \rho_s \tag{8}$$

and the mass flux

$$j = \rho_n v_n + \rho_s v_s \tag{9}$$

Evidence that the viscosity of the superfluid is truly zero is dramatically illustrated by the use of a superfluid gyroscope (Figure 6). Here a toroidal flow channel is packed with jeweler's rouge which enhances superflow. The entire apparatus is set in rotation above T_λ and gradually cooled while rotating. At some $T < T_\lambda$ the rotation is stopped but the superfluid continues to rotate. This is best seen by using the toroid as a fluid gyroscope-any attempt to tilt the channel is accompanied by a precessional motion. Such experiments show that superflows can persist indefinitely. Thermodynamic evidence shows, in addition, that the entropy, S, of helium II belongs entirely to the normal component; the entropy of the superfluid is zero.

There is in addition to the effects we have been discussing the *fountain pressure* (Figure 7a). Two volumes of liquid helium connected by a fine capillary exhibit the unusual behavior of raising the pressure when the temperature is increased on one side. The connection between the pressure and temperature gradients is

$$\frac{\Delta P}{\Delta T} = \rho S \tag{10}$$

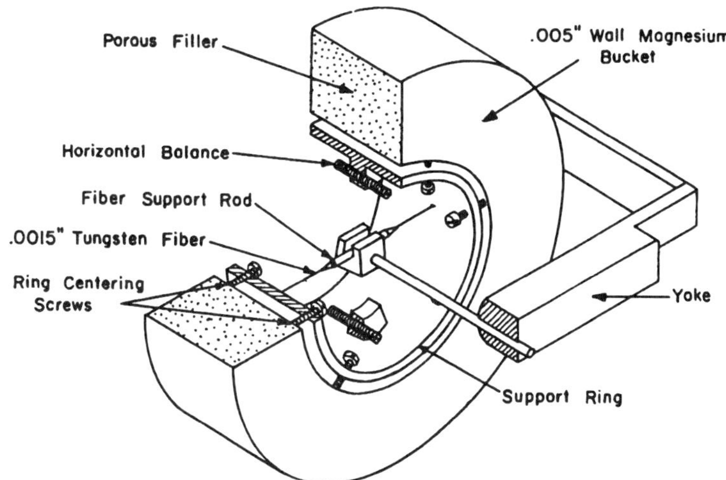

Figure 6 A superfluid gyroscope developed by Reppy's group at Cornell University. The porous filler is used to raise the characteristic velocity of superfluid circulation and the angular momentum of that circulating flow is measured by its gyroscopic effects. (After Kukich, 1970).

The classic device for measuring ρ_n is shown in Figure 8: the Andronikashvili pile-of-disks experiment. The torsion pendulum shown consists of a pile of discs with spacing small compared to the viscous penetration depth at the frequency of the pendulum. The moment of inertia of the pendulum consists of the disc assembly plus the entrained normal fluid. The period of oscillation is measured as a function of temperature and the results determine ρ_n and hence ρ_s from Eq. (6). The results for ρ_n/ρ and ρ_s/ρ found by this and other methods are shown in Figure 9.

An unusual flow peculiar to helium II is illustrated in Figure 10. A channel, heated at one end and cooled at the other end draws superfluid to the heater, and since there is no net mass flux j, Equation (9) shows that normal fluid must counterflow to conserve mass. A vigorous submerged jet can be observed coming from the cool end of the channel, which, however, has no moving parts.

(a)

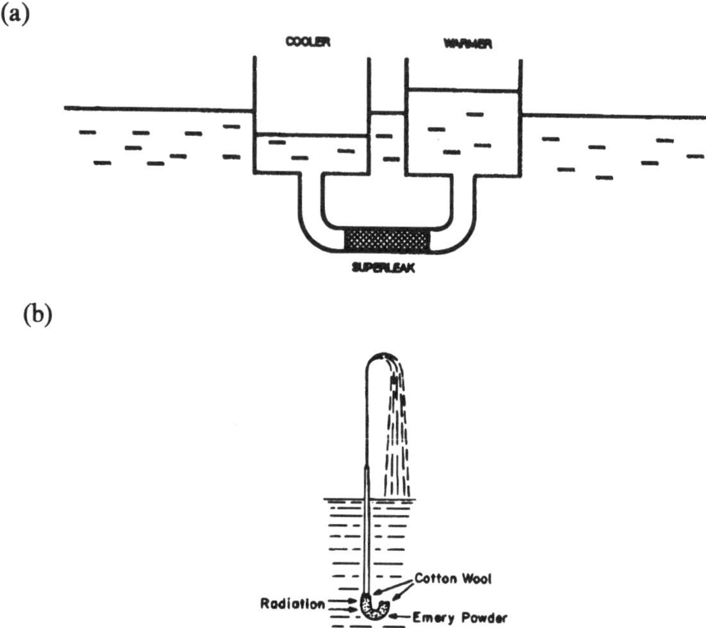

(b)

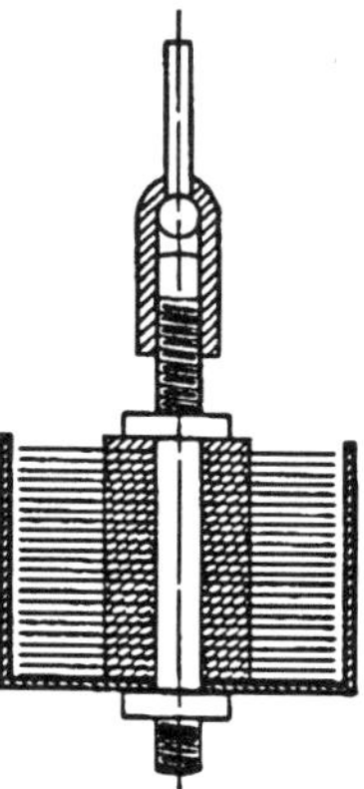

Figure 7 (a) A hypothetical experiment demonstrating the behavior of two vessels of He II connected by a superleak.
(b) A helium fountain making use of the fact that the emery powder immobilizes the normal fluid, and the superfluid flowing toward a source of heat, overshoots to produce a vigorous jet.

Figure 8 The pendulum used in the Andronikashvili pile-of-disks experiment.

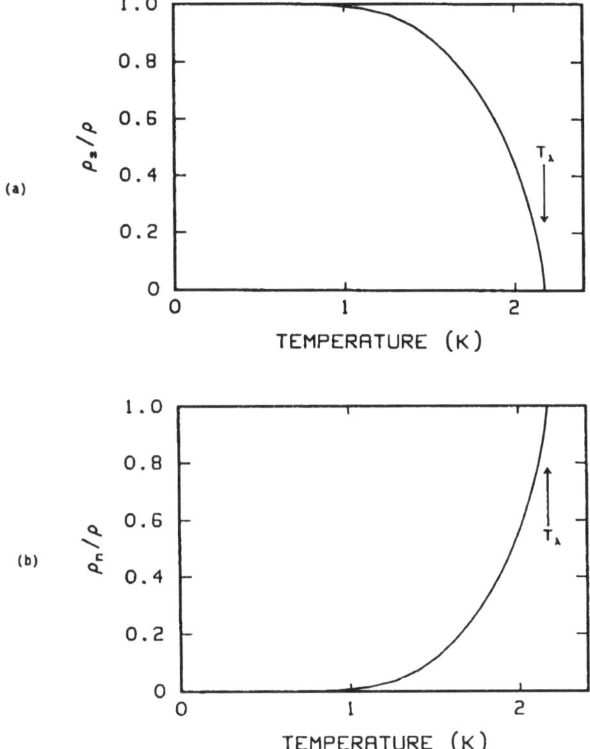

Figure 9 Temperature dependence of (a) ρ_n/ρ and (b) ρ_s/ρ

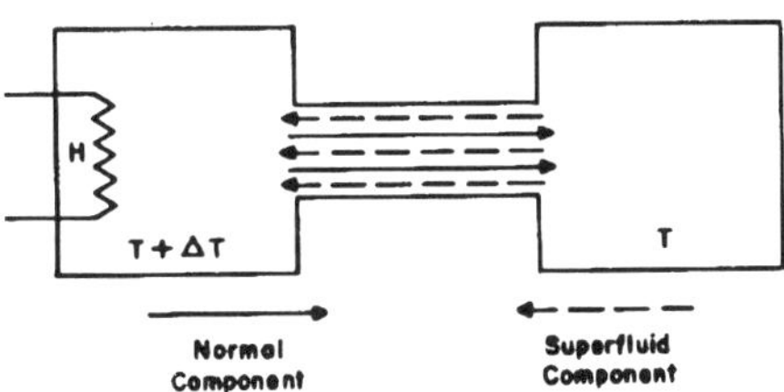

Figure 10 Closed tube containing helium heated at one end and in contact with a heat
reservoir at the other.

If the heater in Figure 10 is switched on and off periodically, the two fluids will set up a periodic reversing counterflow that amounts to a longitudinal standing wave of *second sound*. Second sound is a direct consequence of
the two-fluid model and consists of temperature or entropy fluctuations (as

distinct from density fluctuations for first sound). Second sound reaches velocities as high as ~20m/sec at some temperatures. There can be second sound shock waves by analogy to ordinary shock waves in helium II.

The two fluid model has equations of motion which can be written for small velocities.

$$\rho_s \frac{D v_s}{D t} = -\frac{\rho_s}{\rho}\nabla p + \rho_s S\nabla T \tag{11}$$

$$\rho_n \frac{D v_n}{D t} = -\frac{\rho_n}{\rho}\nabla p - \rho_s S\nabla T + \eta\nabla^2 v_n \tag{12}$$

$$\mathrm{curl}\ v_s = 0 \tag{13}$$

Equation (11) is an Euler equation for the superfluid, (12) a Navier-Stokes equation for the normal component. The terms $\rho_s S\nabla T$ represent the fountain pressure. When $v_n = v_s$ (11) and (12) add up to a Navier-Stokes Stokes equation for the total fluid. Ordinarily $v_n \neq v_s$ because of the irrotational restriction (13) which was put forward by Landau. It denies vorticity to the superfluid.

Experiments by various groups in the 1950s soon showed that (11-13) cannot be complete. The solution for steady rotation in a bucket, for example, shows that the depth of the meniscus must be ρ_n/ρ as deep as for a classical fluid. In practice, however, the meniscus is always its full classical depth.

The solution to this quandary turned out to involve completely new physics: the existence of quantized vortex lines. Figure 11 shows a rotating bucket with an array of quantized vortices, each having circulation $\kappa = h/m = 9.97 \times 10^{-4}$ cm^2/sec and arranged to give the superfluid the vorticity of classical rotation: $n_0\kappa = 2\Omega$ where n_0 is the area density of vortices. The density n_0 is about 2000 lines/cm^2/radian per second rotation rate. Thus the spacing between vortices is about 10^{-2} cm. The size of the core is about 1 Å.

Early experiments also revealed that the counterflow shown in Figure 10 soon fills with a homogenous tangle of quantized vortices above some threshold heat flux.

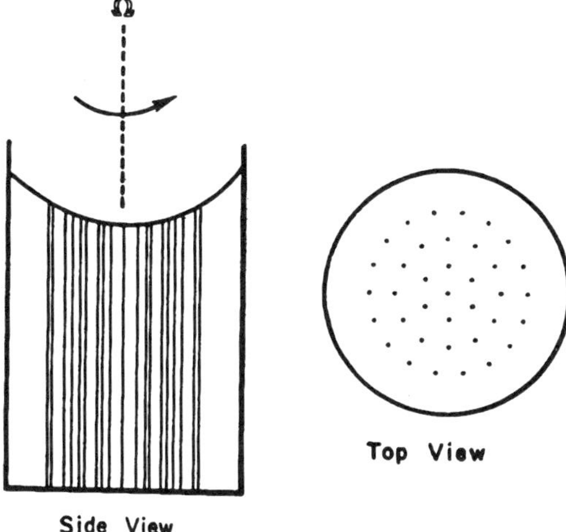

Figure 11 Array of quantized vortices in a rotating bucket. Some vortices are missing near the outer edge.

In either case, once vortices appear, they provide a mechanism to couple the two fluids together. This is called "mutual friction" and was first studied by Hall and Vinen at Cambridge in the 1950s. With mutual friction (13) is dropped and (11) and (12) become

$$\rho_s \frac{D v_s}{D t} = -\frac{\rho_s}{\rho} \nabla p + \rho_s S \nabla T - F_{ns} \qquad (14)$$

$$\rho_n \frac{D v_n}{D t} = -\frac{\rho_n}{\rho} \nabla p - \rho_s S \nabla T + F_{ns} + \eta \nabla^2 v_n \qquad (15)$$

Again, if $v_n = v_s$ these equations add up to a Navier-Stokes Stokes equation for the total fluid. A great deal is known about F_{ns} , but the details need not concern us here. (Donnelly, 1991)

4. Measurement of Superfluid Vorticity: Second Sound and Ion Trapping in Helium II

Second sound (a longitudinal wave) is greatly attenuated by the presence of quantized vortices. For the rotating bucket of Figure 11 propagation across the bucket is strongly attenuated, propagation parallel to the axis is scarcely affected. Hall and Vinen in the 1950s showed that the attenuation coefficient for second sound in rotating helium is

$$\alpha = B\Omega/2u_2 \qquad (16)$$

16   Russell J. Donnelly

where u_2 is the velocity of second sound, and B is a temperature-dependent
coefficient of order unity which must be measured or obtained from theory.
In such an experiment the vorticity $\omega = 2\Omega$ and the vortex line density
is $n_0 = 2\Omega/\kappa$ cm^{-2}, but n_0 can also be thought of as the line density L in
cm/cm^3 of fluid. The result (16) comes from assuming that the mutual friction term F_{ns} in (14) and (15) can be written

$$F_{ns} = -B(\rho_s \rho_n \omega/2\rho)(v_n - v_s) \tag{17}$$

since the vorticity $\omega = \kappa L = 2\Omega$ For a turbulent flow such as in a counterflow channel the equivalent assumption is that

$$F_{ns} = -B\delta(\rho_n \rho_s \omega/2\rho)(v_n - v_s) \tag{18}$$

where again $\omega = \kappa L$ is the magnitude of the vorticity and δ is a constant
of order unity which depends on the details of the flow and its measurement.
In such flows, values of L of 0 - 200,000 cm^{-2} can be generated and the least
line density resolution one can currently achieve is about 20 cm^{-2} corresponds to detecting a change in vortex core volume of about one part in 10^{14}.

Negative ions will stick to quantized vortices at temperatures below ~ 1.7
K. In an array of vortices in rotating helium such as in Figure 11, a beam
of ions produced by an α-emitting radioactive source will be attenuated passing across the bucket. The missing ions can be moved up the vortex lines by
electric fields and collected at the top. Ions have also been used by Schwarz
and his colleagues at IBM to measure the density L of vortex line in a turbulent counterflow. Various strategies can be used to obtain spatial information on the line density.

We therefore have the remarkable situation that the magnitude of the
vorticity can be measured directly in helium II, in contrast to the situation in
classical fluids.

5. Vortex Coupled Superfluidity

The relatively exotic two fluid behavior of liquid helium, described by
Equations 14 and 15 breaks down when the relative velocity $(v_n - v_s)$
becomes large. For many experiments "large" corresponds to a suitably
defined Reynolds number exceeding something of order 100. The form of
the Reynolds number is familiar:

$$Re = UL\rho/\eta \tag{19}$$

but the density appearing in the expression is the total density rather than the
normal fluid density. The evidence for this statement comes from a number
of different experiments, which we summarize briefly.

(a) Rotating Bucket

Historically it was realized that the equations of motion (11-13) would result in motion of only the normal component in a rotating bucket. It is now known that the two fluids rotate together because of the quantized vortices illustrated in Figure 11 and the equations of motion (14-15). As we have explained in Section 4, second sound and superfluidity still exist, but the two fluids are coupled together by vortices. One might call such a state where superfluidity still exists, but vortices are present and modifying the hydrodynamics *vortex coupled superfluidity*.

(b) Oscillating Pile of Disks

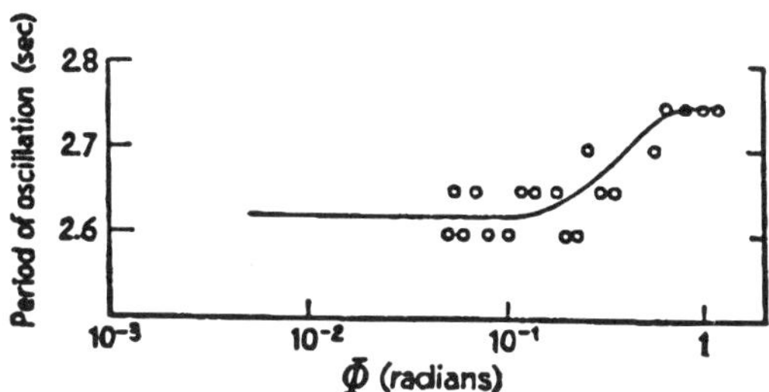

Figure 12 The period of oscillation of a pile of disks at amplitudes below 0.1 radians corresponds to entrainment of the normal component alone. At higher amplitudes, the period corresponds to the entrainment of the entire fluid. (Hollis Hallett, 1955)

We show in Figure 8 the classic Andronikashvili pile of disks. At low amplitudes of oscillation, the superfluid remains undisturbed. We see in Figure 12 that at higher amplitudes of oscillation (above 0.2 radians), the period of oscillation increases and analysis shows that the moment of inertia of the fluid approaches that of the total fluid. Somehow vortices produced at larger amplitudes of oscillation lock the two fluids together. The Reynolds number (19) for such coupling is of order 10, where L is taken as the penetration depth for viscous waves near an oscillating surface. The same physics can be seen in the damping of oscillating disks, spheres and U-tubes and has been reviewed by Donnelly and Hollis Hallett (1958).

(c) Vortex Rings

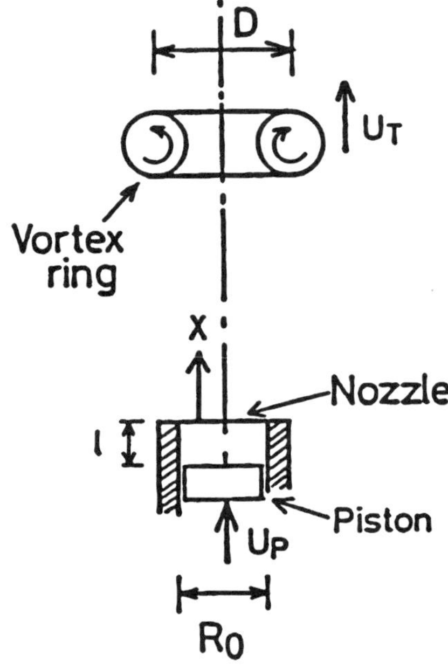

Figure 13 Apparatus constructed by Murakami and Hanada (1988) to study large scale vortex rings in He II, see also Borner et al (1983).

Figure 13 shows an apparatus built by Murakami and his students in Japan to generate large scale vortex rings in helium II by pushing a piston in the generator ejecting helium II through nozzles of 8mm and 14mm. The velocity of the piston up can be varied from 8 to 20 cm/s and the stroke from 5 to 15 mm. Tracer particles are produced by injecting a neutrally buoyant mixture of hydrogen and deuterium gas $H_2 - D_2$. Light came from a 300W xenon lamp providing sheet lighting on the plane of symmetry. The Reynolds number was set between 10^4 and 9×10^4. The results were in accord with behavior of turbulent vortices in a classical fluid as investigated by Tony Maxworthy (1974).

The reader should not confuse these vortices with quantized vortices generated by ions below 1 K (see, for example, Donnelly, 1991).

(d) Thermal Counterflow Jet

Figure 14 shows a apparatus constructed by Murakami and his students to produce a counterflow jet. The heater attracts superfluid which turns into normal fluid and rushes out of the jet, entraining all the fluid with it.

Figure 15 shows flow visualization at four temperatures in helium II and one temperature (2.2K) in helium I at Reynolds numbers from 1600 to 100,000 using neutrally buoyant hydrogen-deuterium mixtures, H_2-D_2, and glass microspheres for visualization.

Figure 16 shows LDV velocities at the exit of a nozzle measured in liquid nitrogen and helium II with a bellows produced flow. These results are especially encouraging, showing particle tracking helium II works exactly as with LN_2.

Figure 17 shows the power spectrum of velocity fluctuations in the helium jet at 1.78K in helium II. The classical Kolmogoroff -5/3 power law is clearly evident.

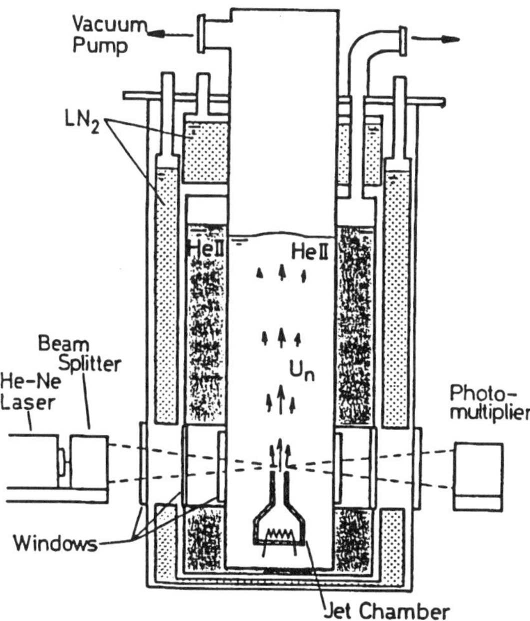

Figure 14 Schematic illustration of the experimental cryostat used by Murakami and his group (Yamazaki and Murakami, 1987).

2-1 : $T = 1.99$ K, $q = 1.06 \times 10^3$ W/m^2, $Re = 1,600$, $Re_n = 500$, H$_2$-D$_2$ particles, 1/2 sec.

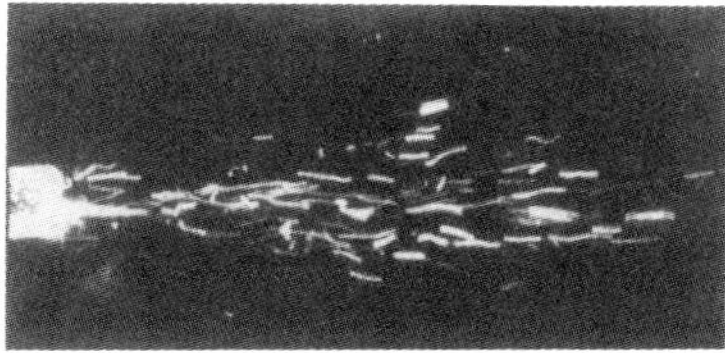

2-4 : $T = 1.63$ K, $q = 1.68 \times 10^4$ W/m^2, $Re = 100,000$, $Re_n = 19,000$, Glass spheres, 1/8 sec.

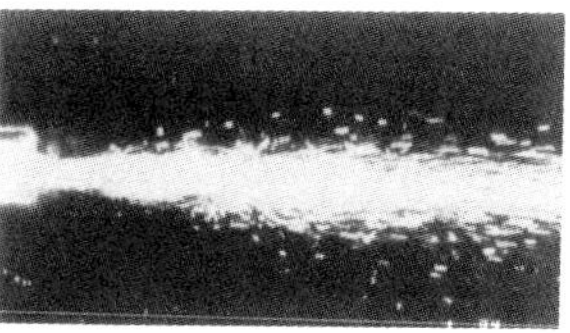

2-2 : $T = 2.12$ K, $q = 1.73 \times 10^4$ W/m^2, $Re = 12,000$, $Re_\perp = 10,000$, Glass spheres, 1/4 sec.

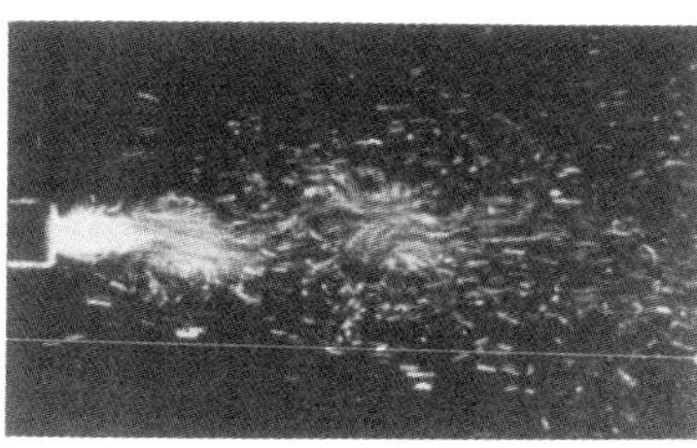

2-5 : $T = 1.90$ K, $q = 3.78 \times 10^4$ W/m^2, $Re = 81,000$, $Re_n = 34,000$, Glass spheres, 1/8 sec.

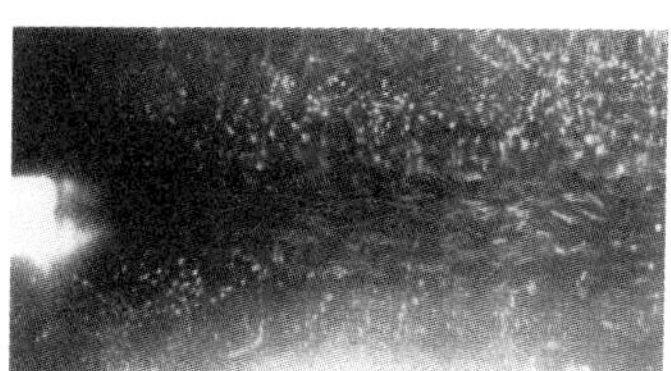

2-3 : $T = 1.95$ K, $q = 1.77 \times 10^4$ W/m^2, $Re = 30,000$, $Re_n = 15,000$, H$_2$-D$_2$ particles, 1/2 sec.

Figure 15 Visualization of the flow from a counterflow jet from Murakami's group. (Murakami and Ichikawa 1987).

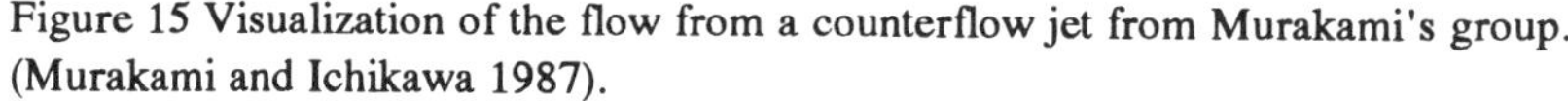

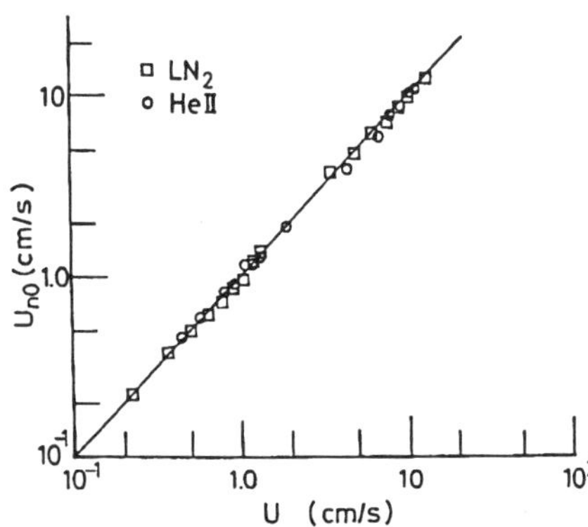

Figure 16 Comparison of LDV results measured at the nozzle with a calibrated flow velocity U produced by a bellows pump. (Yamazaki and Murakami, 1987)

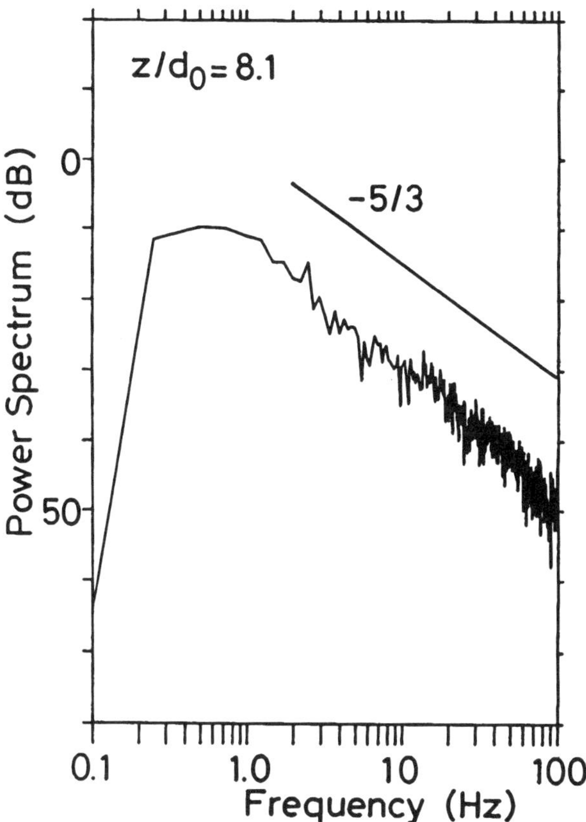

Figure 17 Power spectrum of fluctuations from the counterflow jet. (Yamazaki, 1989)

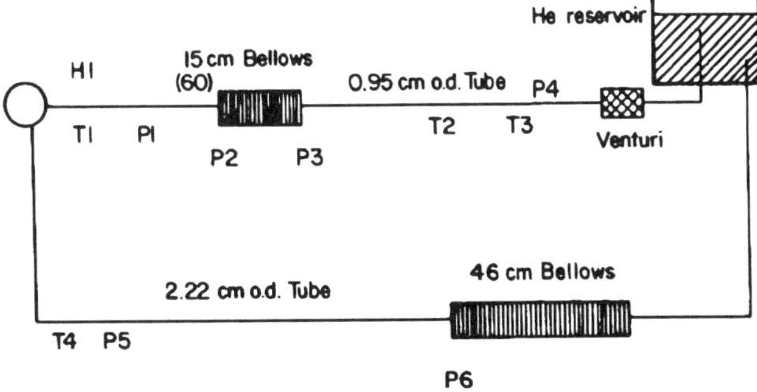

Figure 18 Flow circuit used by Walstrom et al. (1988) to measure the drag of helium II
in a circular pipe. The locations of pressure and temperature sensors are shown.

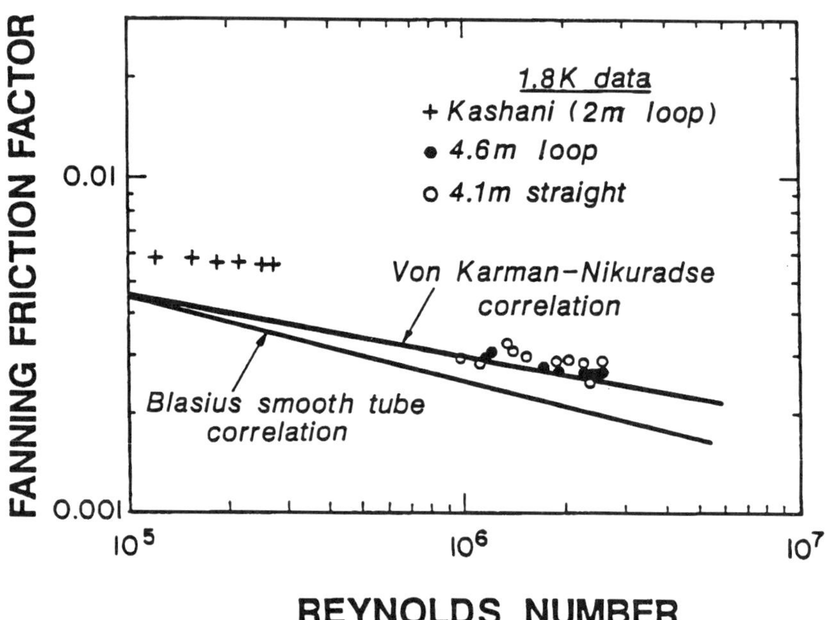

Figure 19 Friction factor for a smooth pipe in helium II. The solid line is the classical correlation, the crosses represent data from a rough pipe. (Walstrom et al 1988).

(e) Friction Coefficient in a Smooth Pipe

Figure 18 shows the apparatus, and Figure 19 the results, of an investigation of the friction factor for helium II obtained by Van Sciver's group at Wisconsin using smooth and rough tubes.

All pressure drop data were converted to Fanning friction factor values f according to the formula

$$f = \frac{\pi^2}{32} \frac{D^5 \rho}{\dot{m}^2} \left(\frac{\Delta P}{L} \right) \tag{20}$$

where D is the diameter, $\left(\frac{\Delta P}{L} \right)$ the pressure gradient and $\dot{m}$ is the mass flow rate. The Reynolds number is $Re = \rho v \, D/\eta = 4\dot{m}/\pi D \eta$.

The results for the smooth tubes are in good agreement with the classical
Von Karaman-Nikuradse correlation up to $Re = 2 \times 10^6$.

The smooth tube correlation is

$$f^{-\frac{1}{2}} = 1.737 \ln\left(Re\ f^{\frac{1}{2}} \right) - 0.396 \tag{21}$$

The rough wall results lie higher as they do in classical flows. Details are
given by Van Sciver elsewhere in this volume.

(f) Drag on a sphere

The drag on a sphere as a function of Reynolds number is one of the classical results in fluid dynamics. [It is shown by the solid curve in Figure 23
below.] It is only natural that the question should arise as to how this curve
looks in helium II. This problem was taken up experimentally by two
groups in the late 1950s. Data at low Reynolds numbers were obtained by
Dowley, Firth and Hollis Hallett (1958, 1961) using a pair of identical
spheres mounted on a torsion suspension in a rotating annulus as illustrated
in Figure 20 (Dowley, 1959).

The Toronto group measured the drag-Reynolds number curve for classical fluids such as CS_2 and found that it did not coincide with the results for a
sphere in a classical wind tunnel. This is perhaps not too surprising, since
rotation introduces completely new physics owing to Coriolis forces and a
further dimensionless quantity, perhaps the Ekman number, would be
required to characterize the results. Nevertheless the principal result is that
there is no marked difference between the results of drag experiments on
helium I or helium II beyond a Reynolds number of several thousand.

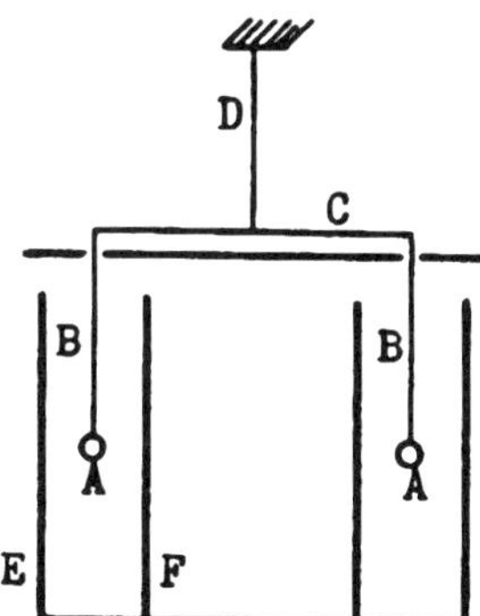

Figure 20 Rotating apparatus used by Dowley, Firth and Hollis Hallett (1958, 1961) to
measure the drag on a pair of spheres mounted on a torsion suspension.

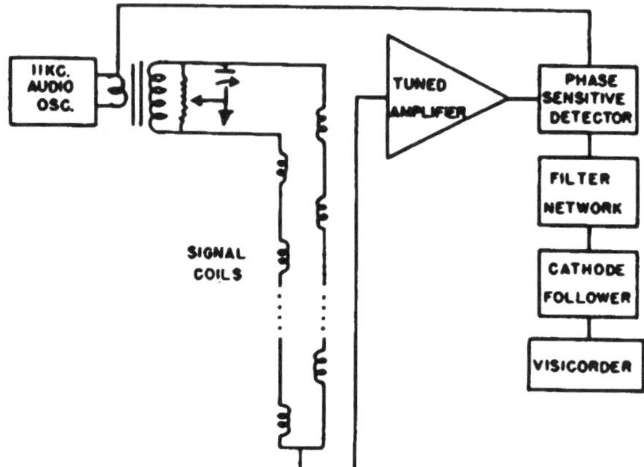

Figure 21 Apparatus used by Laing and Rorschach (1961 a,b) to measure the drag on a falling sphere in helium II.

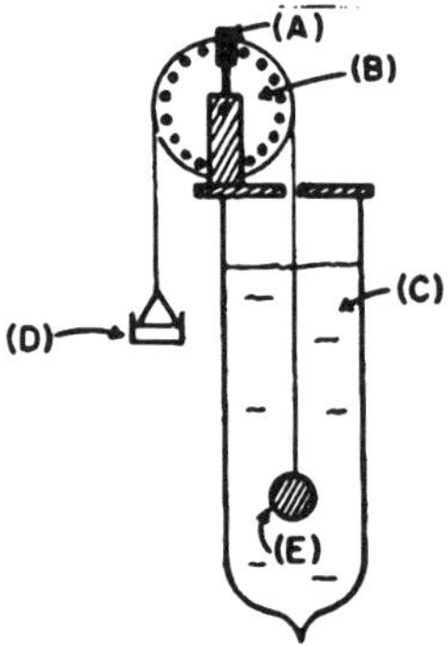

Figure 22 Schematic diagram of the pulley apparatus used by Laing and Rorschach (1961b). A - Phototransistor and light; B - pulley with holes; C - liquid bath; D - counterbalance weight; E - sphere attached to nylon thread.

A further experiment, free of the criticism of rotation, was performed at Rice University by Laing and Rorschach (1961, a,b). Here the drag on a sphere was measured as it fell freely under gravity (Figure 21) or by means of a pulley apparatus shown in Figure 22.

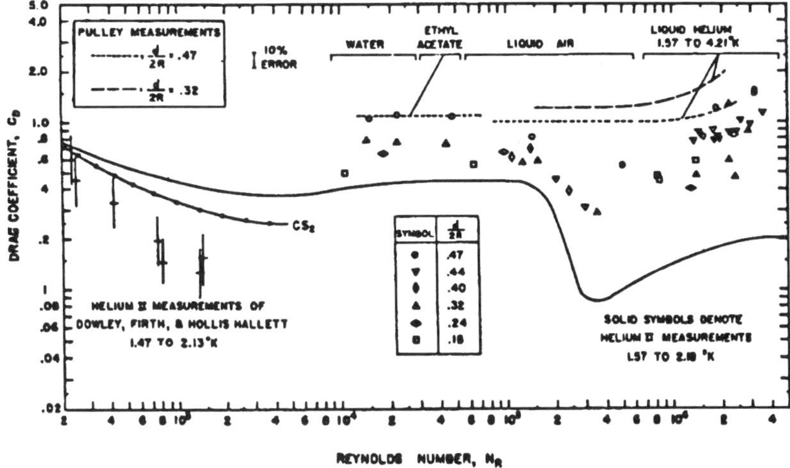

Figure 23 Drag coefficient for a sphere as a function of Reynolds number and sphere diameter. (Laing and Rorschach, 1961b). The solid line is the classical result. The open and solid symbols and the dashed curves are the data obtained by Laing and Rorschach for $10^4 < \text{Re} < 4 \times 10^6$. The results of Dowley, Firth and Hollis Hallett in helium II and CS_2 for $\text{Re} < 5 \times 10^3$ are also shown.

A number of spheres of different density and diameter were dropped in a dewar equipped with twenty evenly spaced coils which detected the sphere as it fell through the liquid in the Dewar flask (Figure 21). A time record was created which allowed the terminal velocity of the sphere to be determined. The apparatus in Figure 22 used an unbalanced pulley system to extend measurement to lower Reynolds numbers. Here the rotation of the pulley was reached by means of light flashes through holes in the pulley.

The results both of the Toronto and Rice experiments are shown in Figure 23. The solid curve is the classical result for drag on a sphere. The curve marked CS_2 is the result obtained at Toronto with carbon disulphide in the same rotating apparatus as was used for liquid helium.

The dotted and dashed curves were obtained at Rice by the pulley method. A substantial wall correction was needed to interpret the results. Laing and Rorschach conclude their analysis with two points:

"Within the rather large scatter, there does not seem to be any difference between the drag on a sphere moving through helium I and one moving through helium II. The hydrodynamic drag properties of helium I and helium II are the same at these high Reynolds numbers"..."The drag crisis does not occur within the range of our measurements...the absence of the drag crisis in liquid helium may be due to the small value of the kinematic viscosity. Since our measurement at a Reynolds number of 10^6 were made over a time interval of less than 0.3 sec, it seems likely that the boundary layer in helium did not have time to reach its steady state thickness...It appears that helium I and helium II behave as ordinary fluids at the high Reynolds numbers of these experiments as far as hydrodynamic drag is concerned. This condition is consistent with the conclusions of Donnelly and Hollis Hallett (1958) drawn from measurements at lower Reynolds numbers...Any equation and boundary conditions proposed for the two fluids must reduce to a single equation of (the Navier-Stokes) form at high Reynolds numbers."

Dowley, in the abstract of his (1959) thesis notes

"in order to obtain a Reynolds number similar to that found in helium I it was necessary to choose a density for the liquid intermediate between ρ and ρ_n ...of the form $\rho_E = \rho_n + f(\upsilon)\rho_s$, where $f(\upsilon)$ is a function of velocity which can vary between zero and unity...The conclusions reached are, that at high Reynolds numbers helium II tends to flow as a normal fluid and conform to the universal (drag) curve, while for low Reynolds numbers the flow of the liquid past the sphere will be different from that of normal fluid at the same Reynolds numbers."

The history of past work in helium II discussed above suggests that for isothermal flows at high Reynolds numbers the two fluids couple together in a state which is hydrodynamically similar to a Navier-Stokes fluid, but still exhibits superfluidity. As we have suggested, this state might be called *vortex coupled superfluidity*.

The work of Laing and Rorschach and Van Sciver suggests that the high Reynolds number vortex-coupled superfluid not only seems to obey the Navier-Stokes equation but also the boundary conditions for smooth and rough surfaces in turbulent flow.

The question then is: what experiments and theories exist or could be carried out to elucidate the transition to and characteristics of vortex-coupled superfluidity? Perhaps the most important theoretical question is to decide exactly how the two fluids couple together in the presence of quantized vortices; in particular how much relative velocity is needed for establishing complete coupling. The empirical equation of Dowley and Hollis Hallett seems to describe the dependence of effective density on velocity. Examples of flow to be examined theoretically might include:

1) Stability of flow between rotating cylinders (Taylor Couette flow)
2) Flow near an oscillating plane
3) Stability of flow in a boundary layer (Tollmein-Schlichting waves)
4) Flow in a circular tube

Theoretical solution of some of these problems might be very difficult indeed, in which case one should explore the option of numerical simulation.

There must be circumstances where vortex coupled superfluidity cannot be entirely classical: for example, second sound propagates in it. Application of a temperature gradient is bound to superimpose a counterflow. All of this needs to be understood.

There is an interesting parallel to vortex coupled superfluidity when the height d of a convecting layer of a dilute solution of ^{3}He in superfluid ^{4}He is much greater than a characteristic dissipation length λ_σ The superfluid mixture behaves as if it were a *single component* classical fluid, but one with the extraordinarily broad and interesting Prandtl number range of $0.04 < Pr < 1.5$. (Metcalfe and Behringer, 1990).

6. Historical Notes on Superfluid Wind Tunnels

Helium was first liquefied in 1908 and by the 1930s efforts were being made in Leiden, Cambridge and Toronto to measure the viscosity. All investigators were aware that low viscosity could lead to high Reynolds numbers and the recording of anomalously high values of viscosity owing to turbulence.

During my student days at Yale in the early 1950s Lars Onsager spoke of a superfluid wind tunnel and suggested the use of superleaks to arrest the flow of the normal fluid and study the pure potential flow of the superfluid. Onsager, of course, was the first to suggest the quantization of circulation for the superfluid.

In 1957, Craig and Pellam, working closely with Feynman at Cal Tech built a small superfluid wind tunnel and demonstrated potential flow for the superfluid. Their apparatus and results are shown in Figure 24.

The superfluid wind tunnel (Figure 24 (a)) consists of a cylindrical region R formed by an evacuated chamber V. R is equipped with two porous plugs S and S' and a heater H. When the apparatus is immersed in helium II, the heater draws superfluid toward it which is evaporated in the pump tube T. Thus in the chamber R, the normal fluid is immobilized and the superfluid moves past the tiny aerofoils W which are suspended by a torsion fiber F. Damping magnets M suppress oscillations of the aerofoil.

Measurements consisted of observing the angular deflection of the suspension and computing the torque as a function of heater power and hence the velocity of the superfluid. The results are shown in Figure 24 (b) and contrast sharply with the expected curve for classical flow. A region of nearly zero lift is observed which verifies the potential flow of the superfluid at low velocities. At velocities above 6 mm/sec quantized vortices begin to couple both velocity fields.

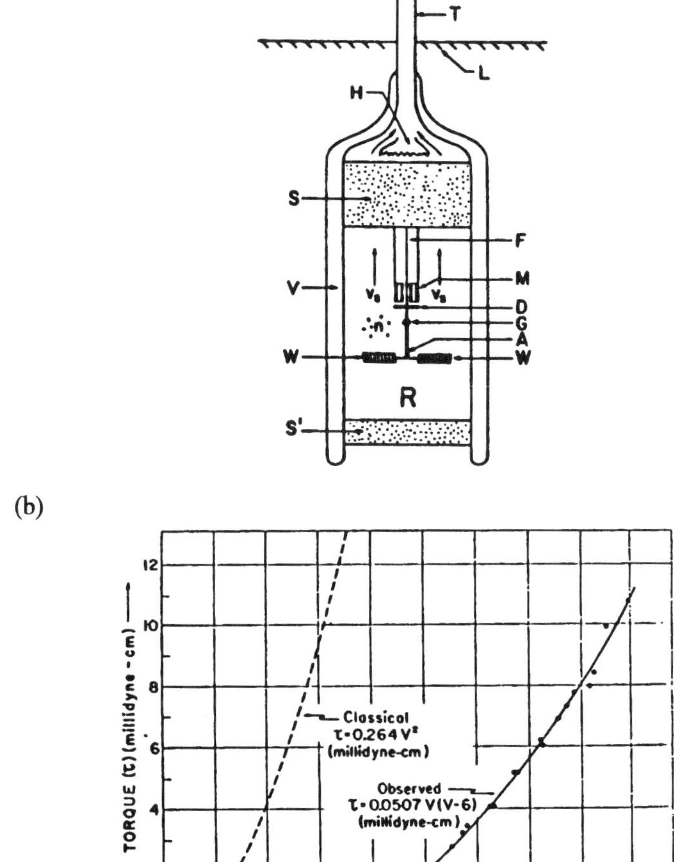

Figure 24 (a) Superfluid wind tunnel apparatus of Craig and Pellam (1957).
(b) Torque on the wings as a function of flow velocity.

On 11-12 January, 1979, a workshop was held at the California Institute of Technology to discuss a proposal by H.W. Liepmann to build a liquid-helium tunnel capable of operating at supersonic speeds. The tunnel test section would be about 3 cm on a side and the operating time would be about 10 seconds in a blowdown mode. Part of the discussion centered around the opportunity provided by such a facility to study the physics and fluid mechanics of liquid helium (a) near the λ transition and (b) near the critical point. Another part of the discussion dealt with instrumentation problems. The remaining discussion centered around costs and benefits of fluid-mechanics research at very high Reynolds numbers which would automatically be accessible in such a cryogenic facility. Since the proceedings of the Cal Tech conference are not widely available, we reproduce here some selected comments by participants. These comments are, of course, quoted out of context.

Hans Liepmann:

> "What we have in mind at GALCIT is not a facility to do aerodynamic testing, but a facility which opens up a new range of parameters in fluid physics.

> Because building a large cryogenic tunnel is a lengthy and expensive undertaking, it seems to us that the scientific community has a responsibility to contribute at an early stage to understanding fluid flow at large Reynolds numbers.

> We are satisfied that it is quite possible to produce very large Reynolds numbers at supersonic Mach numbers in liquid helium."

> ...Most physicists tend to concentrate on incompressible flows where pressure and viscous forces are balanced, like Poiseuille pipe flow or Stokes sphere drag.

> I obviously have not been talking about high Reynolds numbers, but about some very interesting physics. The working fluid is a mixture of normal fluid and super fluid, showing macroscopic quantum effects.

> To summarize, low temperature means high Reynolds number, but also means a strange new fluid with features like the critical point and the lambda line. Gaining access to such a fluid may be worth the trouble of going to low temperature, where gaining access to high Reynolds numbers alone would not."

> I should also say that we have a different ball game in instrumentation. The trend in experiments is to go to shorter times. We would find a 10-second running time long enough.

> The more I look, the more interesting it gets. Fluid mechanics at low temperatures opens up some unusual opportunities, and provides a prospect of finding some basically new phenomena."

Donald Coles:

> "You should understand that we have not made a design study for a cryogenic facility: we have only begun to prepare for such a study. The most relevant technology for a continuous cryogenic tunnel is the technology of pumps for liquid hydrogen rockets such as the Saturn booster.

> My vision of the proposed facility is that a blowdown technique is the most promising one. For a test section area of 10 cm^2 and veloc-

ity 300 m/sec, the flow rate is 300 l/sec. Operation for 10 seconds involves about 3 m³ of liquid.

Because there are small but significant changes in density, some means is needed to maintain constant pressure and velocity. A flywheel of a few hundred kilograms at a few thousand rpm can store the necessary energy. A variable clutch might be used to control the displacement rate."

Hans Liepmann:
"Neither should fluid mechanics be left to physicists."

Philip Klebanoff (in a letter after the workshop):
"I suggest that serious consideration be given to a low-Mach-number facility, perhaps one with a maximum velocity on the order of 10 m/sec. This will still permit significant high-Reynolds-number turbulence research and, perhaps more important, research on the properties of liquid helium above and below the lambda point."

Early in 1988 it occurred to me that a low speed liquid helium flow facility could achieve very high Reynolds numbers, even in helium I. I contacted the Office of Naval Research who made funds available from the Defence Advanced Research Projects Agency. With this support we began a systematic study of various possibilities and gradually made the acquaintance of the various experts speaking at this conference.

7. An Academic High Reynolds Number Flow Facility

We have used the flow loss design information in the 1966 book by Pope and Harper on low speed wind tunnels to arrive at a conceptual design for a flow facility that could be operated in an academic environment. We took as a requirement for such a design that the facility could be built in a conventional low temperature laboratory without major structural changes. At the same time, the capabilities of the facility must be exciting enough to guarantee its usefulness beyond the design and development stage. It is not difficult to see from the figures in Section 2 that adopting a relatively small scale for the flow is not difficult. On the other hand, the losses in the flow go as the cube of the flow velocity, and therefore the design needs to be developed around the refrigeration requirement at liquid helium temperatures. We have somewhat arbitrarily chosen a round figure of 100 watts at 1.6 K as the maximum continuous cooling power to be expected to be available.

The next major quantity to fix is the total volume of liquid helium needed for the working fluid, including dead space in the cryostat. We take the view that a maximum of perhaps 15000 liters would be appropriate to an academic environment -- anything larger would be beyond the capability of laboratory liquefiers to maintain. Within these limits it is not difficult to conclude that a maximum jet diameter of 30 cm would be feasible, with 20 cm perhaps the lower limit of usable test sections given the constraints on building models. We show in Table 1 operating characteristics for 20, 25 and 30 cm test sections in both continuous and transient modes (see below). We have taken cross sections as square. To give an idea of performance with models we have quoted data for drag on a sphere. Here, briefly, is the way the calculations proceed.

The basic configuration of Figure 25 for the tunnel was assumed. The physical dimensions were all expressed as multiples of the cross-sectional dimension of the test section. Thus this cross-sectional dimension together with a desired flow rate (Reynolds number) and temperature became the principal design parameters.

Pressure losses of the individual tunnel components are, in general, a function of the local dynamic pressure q;

$$q = \frac{1}{2}\rho U^2 \qquad (22)$$

Where U is the local flow velocity. Local losses in a particular component were expressed as a product of the local dynamic pressure multiplied by a non-dimensional "loss coefficient" K which depended only on the dimensions of the component, and local Reynolds number (defined below);

$$\Delta p = \frac{1}{2}\rho U^2 K \qquad (23)$$

These loss coefficients have been measured for a wide variety of tunnel configurations and dimensions, the results being expressed in a series of empirical relations of which those in Pope and Harper are a typical (if somewhat conservative) example. The relations in Pope are generally functionally dependent on such variables as local dimensions of the component, and local Reynolds number:

$$Re = \frac{UD}{\nu} \qquad (24)$$

Where D is the local cross-sectional diameter, and ν is the local kinematic viscosity.

Using conservation of mass, we were then able to reference these local loss coefficients back to the dynamic pressure in the test section. Having thus calculated the loss coefficients (properly referenced back to the test section) for a given configuration and set of design parameters, we added them up to arrive at a total loss coefficient Ko_{total}. This total loss coefficient Ko_{total}, when multiplied by the dynamic pressure in the test section yields the total pressure losses around the tunnel. Multiplying this in turn by the cross-sectional area and velocity (in the test section) gives the total power consumption required to overcome pressure losses and keep the fluid moving:

$$Power = \Delta pUA$$

$$= \frac{1}{2}\rho U^2 Ko_{total} UA$$

$$= \frac{1}{2}\rho U^3 A \; Ko_{total} \tag{25}$$

This however assumes a perfectly efficient pump which is unrealistic. The final expression for the power necessary to drive the tunnel is thus

$$Power = \frac{1}{2\eta}\rho U^3 A Ko_{total} \tag{26}$$

where η is the pump efficiency. We have assumed an efficiency of fifty percent for the purposes of this study.

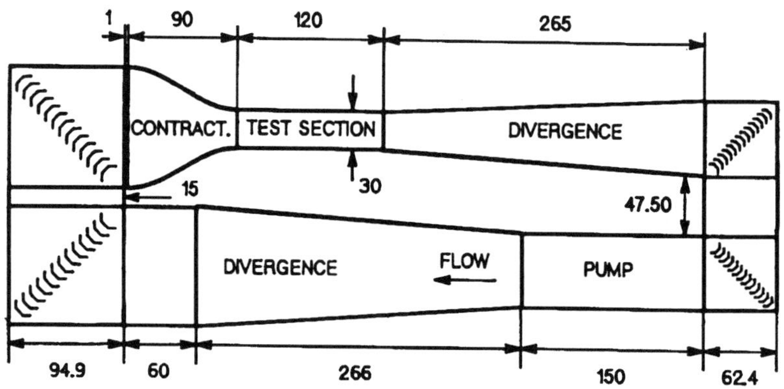

Figure 25 Flow circuit used for the low speed tunnel calculations. The dimensions are in centimeters and the example is for a 30 cm test section.

If we take the most conservative approach, that is to operate in helium I, we still have enormous advantages over water tunnels. For example, supporting the model with a superconducting magnetic suspension system is the *easiest* alternative here. We are well below the transition temperature for many superconductors, and the advantages spelled out in this volume by Lawing can be easily and naturally obtained. Moreover, the sensitivity of such superconducting devices as SQUIDS will automatically be available, extending the power of superconducting technology to wind tunnel testing.

Operation in helium II, however, offers many new possibilities. These include:

(1) Extremely stable temperature control (in small cryostats today, microdegree stability is routine)

(2) Possibility of using a fountain pressure pump to construct a quiet tunnel with no moving parts.

(3) Possibility of direct measurements of vorticity

(4) Unusual flexibility for both pressure and sound measurements

(5) While low velocity measurements are likely to show two fluid behavior, we believe for $R > \sim 1000$ isothermal flow of the vortex coupled superfluid will be classical.

(6) Both liquid and gaseous helium are extremely pure substances, since everything which could contaminate the fluid is frozen out.

Another advantage of using liquid helium is the operation of the tunnel in transient mode. The limiting parameter in the design is the refrigeration available for continuous operation. However, the properties of liquid helium do not vary terribly fast with temperature and it is possible to accelerate the pump, allowing the liquid to warm up in the process. Given that the flow circuit will likely be surrounded by a cylindrical dewar, the total volume of the liquid has a great deal of enthalpy. We show in Table 1 calculations of reasonable transient run times using this strategy. The run times include time to accelerate to the desired operating Reynolds number and decelerate. The gain in Reynolds number is impressive as can be seen from the data in Table 1 where the unit Reynolds number can be doubled for a run period of three to four minutes. Even higher Reynolds numbers can be achieved for shorter periods.

The tunnels of Table 1 can also be operated in helium I with somewhat different characteristics. Table 2 shows the results for the 25 cm tunnel of Table 1: the achievable Reynolds number is somewhat smaller.

A possible fountain pumped flow tunnel is shown in Figure 26. In its simplest form, the heater draws superfluid through the superleak (micron pore filter material) and the contraction and divergence follow. The quality of flow here should be superb since all residual turbulence and swirl should be completely eliminated by the filter material.

When large amounts of flow are involved, the simple picture just described will no longer hold. This is because the superfluid passing through the filter leaves its entropy behind and causes the temperature to rise on the input side of the filter. Heat exchangers could be designed to overcome this effect.

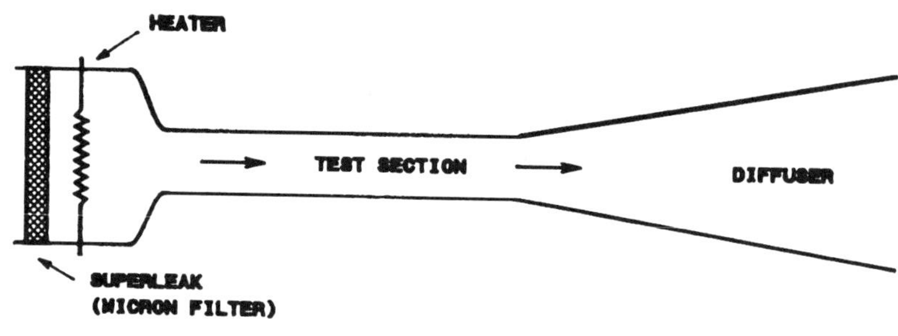

Figure 26 Fountain pump for a flow facility using HeII and having no moving parts.

Table 1

Examples of design of an "academic" high Reynolds number incompressible flow facility using helium II and a refrigerator capable of removing 100 watts. $T = 1.6$ K, $\nu = 9.09 \times 10^{-5}$ cm²/sec			
Size of test section (cm)	20	25	30
Unit Reynolds number (cm^{-1})	5.2×10^6	4.48×10^6	3.98×10^6
Mach number	0.0201	0.0174	0.0154
Flow velocity (cm/sec)	473	408	362
Flow volume (liters/sec)	148.5	200.1	255.7
Dynamic head (lb./ft²)	33.9	25.2	19.8
Shaft power (hp)	0.134	0.134	0.134
Cooling power required (w)	100	100	100
Total power for tunnel (hp)	167	167	167
Total liquid helium (liters)	5310	10,100	17,300
Sphere diameter (cm)	7.14	8.92	10.7
Drag coefficient	0.65	0.65	0.65
Sphere Reynolds number	3.71×10^7	4.0×10^7	4.26×10^7
Transient operation (sec) for unit Reynolds number $= 10^7$ cm^{-1}	184	215	246

Table 2

Operation of a 25 cm tunnel in helium I $T = 2.8$ K, $\nu = 2.6 \times 10^{-4}$ cm²/sec	
Size of test section (cm)	25
Unit Reynolds number (cm⁻¹)	1.56×10^6
Mach number	0.0188
Flow velocity (cm/sec)	405
Flow volume (liters/sec)	199
Dynamic head (lb./ft²)	24.5
Shaft power (hp)	0.135
Cooling power required (w)	101
Total power for tunnel (hp)	95.9
Total liquid helium (liters)	10,100
Sphere diameter (cm)	8.92
Drag coefficient	0.65
Sphere Reynolds number	1.4×10^7

8. Full Sized Test Facilities

Assuming one can work out the details of making reliable tests with liquid and gaseous helium in an academic environment, what sizes of facility might one imagine being made for "national" test installations? We outline three possible designs here: one for incompressible flow such a might be built for testing submarines at high Reynolds number, and two for critical gas tunnels, one for transonic operation and one for low speed high lift studies. We have not taken explicit account of compressibility in these calculations: indeed we used the program outlined in the previous section. When using critical gas, the values of the specific heat ratio γ change with temperature and pressure and hence could change while the gas flows around a model. The exact accounting of these changes is probably a matter of new research. The reader should also understand that the examples given here are adjustable over enormous ranges of parameters and are given only to emphasize the flexibility and utility of helium as a test fluid.

Table 3

Operation of a 125 cm tunnel in helium II $T = 1.6$ K, $v = 9.09 \times 10^{-5}$ cm^2/sec	
Size of test section (cm)	125
Unit Reynolds number (cm^{-1})	3.92×10^6
Mach number	0.0152
Flow velocity (cm/sec)	357
Flow volume (liters/sec)	4377
Dynamic head (lb./ft^2)	19.3
Shaft power (hp)	1.31
Cooling power required (w)	1000
Total power for tunnel (hp)	1630
Total liquid helium (liters)	1.18×10^6
Submarine length (cm)	446
Submarine diameter (cm)	44.6
Drag coefficient	0.10
Submarine Reynolds number	1.75×10^9

Table 3 shows how much could be achieved by running a liquid helium tunnel with 1000 watts of refrigeration instead of 100 as presented in the last section. Here we see that it would be possible to build a 125 cm tunnel for submarine testing which could maintain continuously a Reynolds number on the model of 1.75×10^9. This is high enough for any practical use yet is still a fairly modest facility by standards of large federal installations. For example, the refrigerator for liquid helium at the Continuous Electron Beam Accelerator Facility at Newport News, Virginia operates at 2 K and 4,800 watts.

Now let us turn to full scale critical gas tunnel design. Here we are thinking of tests on aircraft models. The ratio of the specific heats of air $\gamma = 1.4$. Helium gas, being monatomic has $\gamma = 1.66$ in the ideal limit, changing to other values both greater and less than the ideal limit as the critical point is approached and quantum effects become important.

Table 4 shows a design for a low speed, high Reynolds number tunnel for high lift studies. These calculations are based on the same incompressible code used for liquid helium. These helium gas results show several important features. First, we see that very substantial Reynolds numbers based on a chord ten percent of the tunnel diameter can be readily achieved (~70 million). Second, we see that the power to run the fan is essentially negligible, the real power loss is the refrigeration to maintain the tunnel near 6K. The figures assume twenty five percent of Carnot efficiency. Third, we see that the q is very low which means that models could be easily supported by superconducting magnets and would be relatively free of structural deformation at full operating Reynolds numbers. Fourth we note that γ has been increased considerably over the ideal value of 1.66.

Fifth, the pressure ranges are about 1 bar or less and hence do not present unusual construction problems. Finally we observe from Table 4 that desirable operating characteristics can often be achieved by relatively minor changes in temperature and pressure. We show in Figure 4 an example of the rapid changes which occur as the pressure is changed at 5.4 K.

Table 4

Operation of a low speed high Reynolds number tunnel with critical helium gas at three pressures		
Pressure (atm)	0.1	0.5	1.0
Temperature (K)	5.3	5.3	5.3
Size of test section (m)	2.5	2.5	2.5
Mach number	0.3	0.3	0.3
Flow velocity (m/sec)	40.4	39.2	37.6
Re on 25 cm chord, (millions)	7.08	35.3	70.6
Flow volume (Kliters/sec)	198	192	184
Dynamic head q (lb./ft^2)	15.9	79.7	161
Shaft power (hp)	25	114	215
Total power for tunnel (Mw)	4.2	19	36
Gamma	1.69	1.81	2.04

We show in Table 5 a comparison of cryogenic nitrogen and helium tunnels with a heavy gas (SF_6 was one gas for which data was easily available from the manufacturer). We see that assuming 10 atmospheres is about as high as one would want to operate a tunnel for safety reasons, one can reach a chord Reynolds of only 29 million. Heavy gas can reach 70 million by making a very large (4 m) tunnel as shown in Table 6.

Table 5

Comparison of a cryogenic nitrogen high Reynolds number low speed tunnel with a cryogenic tunnel using helium gas and a heavy gas.			
Test fluid	Nitrogen	Helium	SF_6
Pressure (atm)	9.4	1.21	10
Temperature (K)	100	5.4	327
Size of test section (m)	2.5	2.5	2.5
Mach number	0.25	0.25	0.25
Flow velocity (m/sec)	51	31.1	33.8
Re on 25 cm chord, (millions)	70	70	29
Flow volume (Kliters/sec)	250	153	166
Dynamic head q (lb./ft^2)	1120	137	699
Shaft power (Hp)	2040	152	870
Total power for tunnel (Mw)	13.7	25	0.863
Gamma	1.8	2.17	1.13

Table 6

Operation of a 4 m tunnel with SF$_6$ for low speed, high lift. Re = 70 million	
Size of test section (m)	4
Temperature (K)	300
Pressure (atm)	11
Mach number	0.25
Flow velocity (m/sec)	33.8
Flow volume (Kliters/sec)	424
Dynamic head (lb./ft²)	884
Shaft power (hp)	2740
Total power for tunnel (Mw)	2.04

An example of transonic operation is given in Table 7. Achieving a high Mach number is expensive in terms of refrigeration if the desire is to operate continuously. The NTF gets around the refrigeration problem by spraying liquid nitrogen directly into the stream. At top Reynolds number, the NTF uses 250,000 gallons of liquid nitrogen in 30 minutes.

The shaft propulsion requirements are not large, again the bulk of the helium tunnel power is refrigeration. If an intermittent helium gas supply could be designed the cost could be brought down. The problem is that liquid helium itself is the only convenient source of cold at these temperatures. Here the Cal Tech proposal for a blowdown technique would seem to merit renewed consideration.

Table 7

Comparison of a cryogenic nitrogen transonic tunnel with a cryogenic helium gas tunnel and a heavy gas tunnel working near room temperature.			

Test fluid	Nitrogen	Helium	SF_6
Pressure (atm)	3.3	0.38	7.8
Temperature (K)	100	5.4	327
Size of test section (m)	2.5	2.5	2.5
Mach number	0.8	0.8	0.8
Flow velocity (m/sec)	163	107	108
Re on 25 cm chord, (millions)	70	70	70
Flow volume (Kliters/sec)	800	523	530
Dynamic head (lb./ft^2)	3350	434	5470
Shaft power (hp)	19,500	1650	21,100
Total power for tunnel (Mw)	131	270	21
Gamma	1.5	1.77	1.13

9. Free Surface Testing

Liquid helium has properties which often vary rapidly with temperature. In this section we shall see that these properties can be used to model the motion of ships or waves on a free surface in ways not possible with water.

In modelling surface ships we need above all to watch the Froude number

$$Fr = U/(gL)^{1/2} \qquad (27)$$

where L is the length of the vessel. The Froude number characterizes resistance by wave generation. Suppose we define the scale ratio λ as

$$\lambda = L_w/L_h \qquad (28)$$

and adopt the subscript w to denote full scale in water and subscript h for a laboratory model. If we require $Fr_w = Fr_h$ then

$$U_h/U_w = (L_h/L_w)^{1/2} = \lambda^{-1/2} \qquad (29)$$

There are, however, other scaling parameters at a free surface, for example the Weber number

$$We = \rho U^2 L/\sigma \qquad (30)$$

where σ is the surface tension, governs any surface tension effects, the Reynolds number governs turbulence effects

$$Re = UL/\nu \tag{31}$$

and finally the ratio α of density of liquid to density of the gas above it. The parameter α and the ratios

$$\frac{Re_h}{Re_w} = \frac{U_h L_h \nu_w}{U_w L_w \nu_h} = \frac{\nu_w}{\nu_h} \frac{1}{\lambda^{3/2}} = \frac{\beta}{\lambda^{3/2}} \tag{32}$$

$$\frac{We_h}{We_w} = \frac{\rho_h \sigma_w}{\rho_w \sigma_h} \frac{1}{\lambda^2} = \frac{\gamma}{\lambda^2} \tag{33}$$

where

$$\beta = \nu_w/\nu_h, \gamma = \rho_h \sigma_w/\rho_w \sigma_h \tag{34}$$

define the material parameters α, β and γ which are temperature dependent and must be computed for every desired set of operating conditions. One example is given in Table 8. From Table 8 and the equations above one can see that if we test in water where $\beta = \gamma = 1$ we can match Froude numbers by using Equation (29), but now (32) shows that the Reynolds number is too small by a factor $\lambda^{3/2}$ and (33) shows that the the Weber number is too small by a factor λ^2.

To illustrate the situation in current use, let us consider the motion of a surface ship 200m long moving at 32 knots (16.5 m/sec). Let us assume the water temperature is 15C where the kinematic viscosity is 1.141×10^{-2} cm²/sec. The surface tension depends on contaminants and can range over a factor of two from 35 to 70 dynes/cm. We shall take the lower value as consistent with contaminated water rather than pure. If we take $\lambda = 25$ then we have a model 8m long in a water tow tank and the Froude number (0.373) will be matched if the model is towed at 3.3m/sec. However, under these conditions the Reynolds number of the model is 2.34 x 10^7 compared to the full-scale Reynolds number of 2.89 x 10^9, that is, the Reynolds number for the model is a factor of $\lambda^{3/2} = 125$ too small. Similarly the Weber number of the ship is 7.46 x 10^8 and that of the model is 1.19 x 10^6, a factor of $\lambda^2 = 625$ times too small. The water/air density ratio is, of course, correct.

Suppose we consider using a liquid helium tow tank. Consider the tank as a cylinder of radius r and length L_T which is half full. To find the right parameter space, let us first fix the density ratio to be the same as water and air. This leads to an operating temperature of T=1.53K. If we use (32) to require that the Froude number *and* the Reynolds number match then $\lambda^{3/2} = \beta$, or $\lambda = 24.5$ from Table 8. We see that at 1.53K we can match the density ratio, Froude number and Reynolds number, but the Weber number of liquid helium is a factor 21 too small. This is a vast improvement over water, however, for which the Weber number is a factor 627 times too small.

Note that the matching with liquid helium involves picking the scaling ratio λ from the material parameter β, then picking the tow velocity by (29).

The volume of liquid helium to fill this tow tank is $\frac{1}{2}(\pi r^2)L_T$ where L_T is the length of the tank. For $r = 8m$ the volume is

$$V = \frac{1}{2}(\pi r^2)L_T = 101\,L_T\,m^3 \tag{35}$$

The length L_T depends on the desired run time τ. For 10 seconds of run time at 3.3 m/sec, $L_T = 33$ m and $V = 3.32 \times 10^3$ m³ $= 3.32 \times 10^6$ liters of liquid helium.

We could choose instead, for a wave tank for example, to match the density ratio, Froude number and Weber number. Then $\lambda^2 = \gamma$ and $\lambda = 3.91$ and the Reynolds number is 15.7 times too large.

For geophysical fluid dynamics problems, the fluid is in rotation at angular velocity Ω and the Rossby number and Ekman numbers (6) and (7) are significant for rotating systems. The kinematic viscosity of the liquid used for testing is a factor β smaller than the kinematic viscosity of water. This produces a significant reduction in Ekman number for liquid helium.

Plots of the dimensionless material parameters β and γ as functions of temperature are shown in Figures 27 and 28.

Table 8

Modelling Motions on the Air/Water Surface in the Laboratory			
Test Fluid	α	β	γ
Water	1.23×10^{-3}	1	1
Liquid helium 1.53 K	1.21×10^{-3}	121	15.3

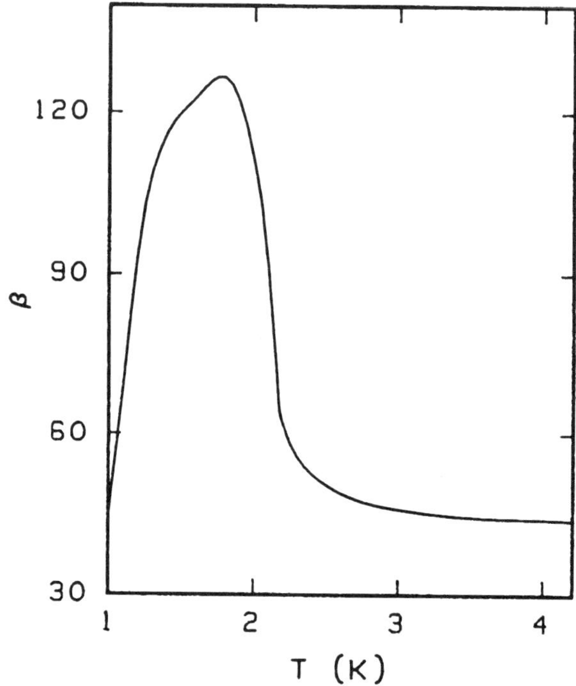

Figure 27. Variation of the parameter β with temperature. The largest values of β and hence of scale parameter λ occur about 1.75K if the desire is to match Reynolds number exactly.

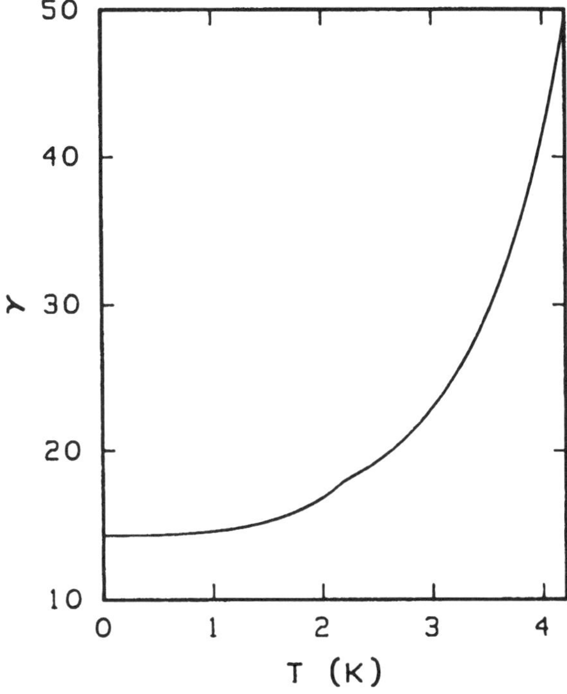

Figure 28. Variation of the parameter γ with temperature. The largest values of γ occur at higher temperatures, giving the largest values of λ if the desire is to match Weber number exactly.

10. Design Considerations for Fountain Pumps

In recent years, a large body of research has been accumulated on the efficiency and design of Fountain Effect Pumps (FEP). This work has been funded largely with an eye to developing an on-orbit liquid helium transfer capability for the Space Shuttle (Snyder, 1988). The most appropriate mode of operation for a FEP in a closed loop wind tunnel such as in Figure 26 here seems to be one of constant heat input to the the pump ($d\dot{Q} = 0$). Kittel (1986, 1987, 1988a, 1988b) develops equations relating Δp and $\dot{m}$ (see figure 26):

$$\frac{6.6\Delta p}{\rho \dot{Q}} = \frac{1}{\dot{m}} - \frac{1}{\dot{m}_0} \tag{36}$$

where

$$\dot{m}_0 = \frac{Q}{S_0 T_0} \quad \text{and} \quad \dot{Q} = \dot{m} S_1 T_1 \tag{37}$$

From simple hydrodynamic considerations, one can determine appropriate expressions for the pressure drop, $\Delta p = p_1 - p_0$ around a closed loop (typically $\Delta p \propto V^2$ where V is the velocity in the test section) and of course the mass flow rate $\dot{m} = \rho V A$. Together with the desired test section operating temperature T_1, this allows one to determine the necessary temperature upstream of the pump.

In properly designing such a pump, there are many things which must be considered, including for example: proper placement of the heater downstream of the superleak, effects of normal fluid flow in the superleak, and conduction of heat through pump assembly to the surrounding fluid. Probably most important for this general discussion however, are the issues of critical velocity within the superleak, and heat management upstream of the pump.

Performance calculations for ideal superfluid pumps are usually founded on the assumption that the fluid velocity within the pump is below the first critical velocity. Putterman (1974) suggests the following empirical relation:

$$v_{s,c} d^{1/4} = 1.0 \, (c.g.s.) \tag{38}$$

where we take d as the characteristic pore size, and $v_{s,c}$ is the critical velocity. With a pore size of approximately 1 micron, $v_{s,c} \sim 10$ cm/sec through the superleak. DiPirro et al (1988) indicate that ceramics with an appropriate pore size are available off the shelf.

11. Studies of High Reynolds Number Turbulence

The field of turbulence is receiving renewed attention in both engineering and in physics. Generally speaking, the limitation on experimental work is the generation of the significantly high Reynolds numbers. Liquid and gaseous helium offer the opportunity to generate unprecedented Reynolds numbers in a strictly controlled environment. An example of such an experiment is discussed by Smith and Donnelly elsewhere in this volume.

Many turbulence experiments are best conducted in a continuous wind tunnel such as we have shown in Table I. The design of such a facility should be as large as possible for turbulence work in order that the smallest eddies of interest can be studied. At Reynolds numbers of order 10^7 the smallest scales will be of order one micron. Modern nanostructures, however, make studies on this scale quite feasible.

An interesting outline of modern problems in turbulence is contained in a recent article by Frisch and Orszag (1990). Some new directions are discussed by Sreenivasan and Schwarz in further chapters.

If a sizeable liquid flow facility were constructed, the storage containers for the fluid could be used to house Bénard convection experiments as described in Section 2. One can extend the maximum Rayleigh number dramatically this way, as it varies as the cube of the height of the convection chamber.

12. Conclusions

We have outlined the use of helium as a test fluid and find that it has many uses which could well be explored in the future. Helium gas is an alternative to air or cooled nitrogen for wind tunnels where very high Reynolds Numbers are desirable. The best designs for tunnels needing considerable Mach effects will likely be of the blowdown variety.

Liquid helium would be useful and economical for incompressible flows at very high Reynolds numbers. Helium II is likely the preferred liquid phase because of its remarkable properties which allow unprecedented temperature stability. The two fluid properties which are so spectacular at low Reynolds numbers appear to be much less important at high Reynolds numbers. Helium I is certain to be useful as it is known to be a classical fluid.

The rapidly varying properties of liquid helium with temperature are likely to make free surface studies important. For conventional tow tank studies liquid helium offers a match of both Reynolds and Froude numbers, overcoming a familiar limitation of conventional tow tank studies of surface vessels.

13. Acknowledgements

This conference and the study of helium wind tunnels was sponsored by the Office of Naval Research through funds from the Defence Advanced Projects Agency. My research in low temperature physics is supported by the National Science Foundation Low Temperature Physics Program.

Many people have contributed to the thinking expressed in this article. In particular Michael Smith carried out all the wind tunnel calculations and fountain pump calculations quoted here as well as carrying out the measurements with the towed grid reported elsewhere in this volume. Charles Swanson did the first critical look at free surface testing. Robert Kilgore and Ronald Smelt were kind enough to read the entire draft with great care.

14.Bibliography

Behringer, R.P. and Ahlers, G. (1982) "Heat transport and temporal evolution of fluid flow near the Rayleigh-Bénard instability in cylindrical containers", J. Fluid Mech. **125**, 219-258.

Borner, H., Schmeling, T., and Schmidt, D.W. (1983) "Experiments on the circulation and propagation of large scale vortex rings in HeII" Phys Fluids **26**, 1410-1416.

Craig, P.P. and Pellam, J.R. (1957) "Observation of perfect potential flow in superfluid", Phys. Rev. **108**, 1109-1112.

DiPirro, M.J., Quinn, E.R. and Boyle, R.F. (1988) "Tests of a nearly ideal, high rate thermomechanical pump", *Proc. 12th International Cryogenic Engineering Conference*, Southampton, UK, 646.

Donnelly, R.J. and Hollis Hallett, A.C. (1958) "Periodic boundary layer experiments in liquid helium", Annals of Physics **3**, 320-345.

Donnelly, R.J. (1967) *Experimental Superfluidity*, Chicago Univ. Press

Donnelly, R.J. (1991) *Quantized Vortices in Helium II*, Cambridge Univ. Press

Dowley, M.W. (1959) *The Drag Coefficient of a Sphere in Liquid Helium* PhD Thesis, University of Toronto (unpublished).

Dowley, M.W., Firth, D.R. and Hollis Hallett, A.C. (1958) "Drag coefficient of a sphere in liquid helium II", Proceedings of the Fifth International Conference on Low Temperature Physics. 19-21.

Dowley, M.W., Firth, D.R. and Hollis Hallett, A.C. (1961) "Drag coefficient of a sphere in liquid helium II", Proceedings of the Seventh International Conference on Low Temperature Physics, 464-465.

Frisch, U. and Orszag, S.A. (1990). "Turbulence: challenges for theory and experiment", Physics Today, January, 24-32.

Hollis Hallett, A.C. (1955) "Oscillating disks and rotating cylinders in liquid helium II" in *Progress in Low Temperature Physics*, Vol. I, ed C.J. Gorter, North-Holland, Amsterdam.

Kittel, P. (1986) "Losses in fountain effect pumps", *Proc. 11th International Cryogenic Engineering Conference*, Butterworth, London, 317-323.

Kittel, P. (1987), "Liquid helium pumps for in-orbit transfer", *Cryogenics*, **27**, 81-87.

Kittel, P. (1988a), "Temperature rise in superfluid helium pumps", NASA TM-100997, July 1988.

Kittel, P. (1988b), "Operating characteristics of isocaloric fountain-effect pumps", *Advances in Cryogenic Engineering*, **33**, 465-470.

Laing, R.A. and Rorschach, H.E. (1961a) "Hydrodynamic drag force on spheres falling in liquid helium", Proceedings of the Seventh International Conference on Low Temperature Physics. 461-463. Physics of Fluids **4** 564-571.

Laing, R.A. and Rorschach, H.E. (1961b) "Hydrodynamic drag on spheres moving in liquid helium", Physics of Fluids **4**, 564-571.

Libchaber, A. (1987) "From chaos to turbulence in Bénard convection" Proc. Roy. Soc. A **413** 63-69.

Maxworthy, Tony (1974) "Turbulent vortex rings", J. Fluid Mech **64**, 227-239.

Metcalfe, G.P. and Behringer, R.P. (1991) "Using superfluid mixtures to probe convective instabilities" Physica D (in press)

Murakami, M. and Ickikawa, N. (1987) "Flow visualization study of thermal counterflow jet in He II" J. Cryogenic Soc. Japan **22**, 175-180.

Murakami, M. and Hanada, M. (1988) "Flow visualization study on large-scale vortex ring in HeII" Proc ICEC 12 Butterworths, Guildford, UK 281-285.

Pope, A. and Harper, J.J. (1966) *Low Speed Wind Tunnel Testing*, John Wiley and Sons, New York [Second Edition: W.H. Rae and A. Pope, John Wiley and Sons, 1984]

Putterman, S.J. (1974) *Superfluid Hydrodynamics*, North-holland/American Elsevier.

Snyder, H.A. (1988) "Dewar to dewar model for superfluid helium transfer", Cryogenics **28**, 86-89.

Smelt, Ronald (1945) "Power economy in high speed wind tunnels by choice of working fluid and temperature", (Reproduced in the Appendix)

Threlfall, D.C. (1975) "Free convection in low-temperature gaseous helium", J. Fluid Mech. **67**, 17-28.

Tritton, D.J. (1988) *Physical Fluid Dynamics*, Second Edition, Clarendon Press, Oxford.

Walstrom, P.L., Weisend II, J.G., Maddocks, J.R. and Van Sciver, S.W. (1988) "Turbulent flow pressure drop in various He II transfer system components" Cyrogenics **28**, 101.

Yamazaki, T. (1989) *Study of Thermal Counterflow Jet in HeII Using a Laser Doppler Velocimeter*, Masters Thesis University of Tsukuba (unpublished).

Yamazaki, T. and Murakami, M. "Application of laser doppler velocimeter to HeII flow - investigation of circular HeII jet", J. Cryogenic Soc. Japan, 49-54

Modern Wind Tunnels

CRYOGENIC WIND TUNNELS

Robert A. Kilgore

NASA Langley Research Center, Hampton, VA 23665-5225

ABSTRACT

This paper opens with a brief review of cryogenic wind tunnels and their use for high Reynolds number testing. Emphasis is on operational and aerodynamic testing experience in the NASA Langley 0.3-m Transonic Cryogenic Tunnel (TCT). Since we built the 0.3-m TCT in 1973, it has logged over 8000 hours of running at cryogenic temperatures. We use the 0.3-m TCT for aerodynamic testing and to develop test techniques for cryogenic tunnels.

Areas briefly covered in this paper include development of test techniques and aerodynamic testing in cryogenic tunnels. Based on our experience, we recommend using advanced testing techniques to increase the value of cryogenic tunnels to the research community. These include **adaptive wall test sections,** using solid but flexible top and bottom walls, **magnetic suspension and balance systems** and **non-intrusive laser techniques..**

INTRODUCTION

The ability to simulate the aerodynamics of flight is constantly improving. One area of improvement is testing at flight values of Reynolds number in subsonic and transonic wind tunnels. The relatively recent development of cryogenic wind tunnels makes it possible to test at flight Reynolds number. This paper reviews the state of development of cryogenic wind tunnels and their use for high Reynolds number testing.

THE WORLD'S CRYOGENIC WIND TUNNELS

I begin by defining what I mean by *cryogenic.* The U.S. National Bureau of Standards considers the field of cryogenics to involve temperatures below 123 K (-150°C). This definition for cryogenics is as good as any. The normal boiling points of the so-called permanent gases (helium, hydrogen, neon, nitrogen, oxygen, and air) all lie below 123 K. Freon refrigerants, hydrogen sulfide, and other common refrigerants all boil above 123 K. Thus, by this definition, there are about 20 cryogenic wind tunnels in use around the world. A review article in *Cryogenics* describes many of these cryogenic tunnels.[1]

Emphasis in the review article is on the cryogenic engineering aspects of the design and operation of cryogenic wind tunnels. For a variety of reasons, some of the earlier cryogenic tunnels have been abandoned or become inactive. These include the 1 foot and 4 foot Blowdown Cryogenic Tunnels built at Douglas and the fan driven T3 at ONERA/CERT.

There are several cryogenic wind tunnels in various stages of planning, design, and construction. One of the more significant in this category is the European Transonic Windtunnel (ETW) which I will describe later.

In this section I briefly describe nine cryogenic tunnels. Each of these tunnels is either of historical interest or of interest because of its actual or potential research capability. These tunnels range from the very simple to the very complex. They are examples of the actual or anticipated use of cryogenic tunnels for high Reynolds number testing.

SUBSONIC TUNNELS

<u>The First.</u> The first cryogenic tunnel was a very simple atmospheric low-speed tunnel. It started as an existing 1/24 scale model of the Langley V/STOL tunnel. We modified this tunnel and successfully operated it as the world's first cryogenic tunnel. It first ran at cryogenic temperatures in

January of 1972. Through the spring and summer of 1972 we used this tunnel for a variety of proof-of-concept tests. Figure 1 shows the circuit of the low-speed cryogenic tunnel which is typical of most fan-driven cryogenic wind tunnels.

One of our major concerns with the low-speed tunnel was finding a simple way to cool the tunnel to cryogenic temperatures. Another concern was finding suitable thermal insulation for the circuit. We had many other concerns. For example, could we really use wooden fan blades in a cryogenic tunnel? With the clarity of hindsight we now see that most of our early concerns about cryogenic tunnels were unfounded. As wind tunnel engineers, we simply had to extend our experience to lower temperatures. The work at Langley on the low speed tunnel let us discover some of the real problems with cryogenic tunnels. It also let us find solutions to many of the problems on a small scale at low cost.

Once we had worked out operating procedures, we used the low-speed tunnel to confirm the validity of the cryogenic concept. Reference 2 gives details of our work with the low-speed tunnel. Table 1 lists the main conclusions we drew from the experiments and from the day-to-day operation of the low-speed tunnel.

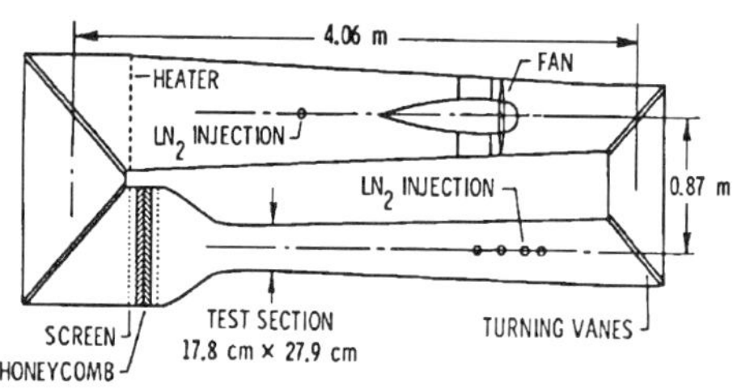

Fig. 1- Sketch of Low-Speed Cryogenic Tunnel

Aerodynamic
* Boundary-layer development with Reynolds number the same for ambient and cryogenic conditions.
* Drive-power and fan-speed decrease as predicted.

Operational
* Cooling with LN$_2$ is practical.
  - Rapid cooldown
  - Automatic temperature control
  - Gas stream is clear, dry, and frost-free
* Can use conventional strain-gage balance.
* Trouble-free operation of drive motor and fan.

Table 1   Low-speed Cryogenic Tunnel Results

Type	closed circuit, fan
Material of construction	concrete
Insulation	internal
Cooling	liquid nitrogen
Test gas	nitrogen
Test section size (h,w,l)	2.4 x 2.4 x 5.4 m
Mach range	up to 0.38
Contraction ratio	10.3:1
Stagnation pressure	up to 1.12 bars
Stagnation temperature	100 - 300 K
Running time	several hours
Max. Reynolds number/m	37 million
Drive motor	1 MW
Fan speed	up to 500 rpm
LN$_2$ tank volume	150 m^3

Table 2   KKK Cryogenic Low-Speed Tunnel

The Biggest. Significant because of its size and its research potential is the low speed cryogenic tunnel of DLR, the Kryo-Kanal-Koeln (KKK).[3]

Viehweger and his co-workers at the DLR Research Center at Porz-Wahn modified a large low-speed tunnel for cryogenic operation. The project started in 1978 with studies of how to modify an existing low-speed 3 m wind tunnel. The studies included modeling the LN$_2$ injection process and finding ways of fixing internal insulation to the concrete tunnel. They completed the studies in 1979 and got approval to continue the project in 1980. The first cryogenic operation of KKK was in January of 1986.

The KKK is a thoroughly modern low-speed cryogenic wind tunnel. The work by Viehweger and his colleagues is an excellent example of how to modify an existing tunnel for cryogenic operation. By careful study, they understood and solved the problems of liquid nitrogen injection, internal insulation, and automatic controls. They now have an excellent low-speed high Reynolds number tunnel.

Reference 3 gives details about the design and operation of KKK. Table 2 gives some of the design and operational characteristics of the KKK.

TRANSONIC TUNNELS

The First. The 0.3-m Transonic Cryogenic Tunnel TCT) at NASA Langley is the world's first transonic cryogenic tunnel. It can operate from ambient to cryogenic temperatures at pressures up to 6 bars. We built and first operated the 0.3-m TCT in 1973 with a 33 cm octagonal 3-D test section. We used the 3-D test section for all of the early experimental proof-of-concept studies.[4]

Type	closed circuit, fan
Material of construction	aluminum
Insulation	external, purged
Cooling	liquid nitrogen
Test gas	nitrogen
Test section size (h,w,l)	33 x 33 x 142 cm
	(solid adaptive walls)
Mach range	0.02 to 1.3 +
Contraction ratio	10.7:1
Stagnation pressure	1.1 - 6.2 bars
Stagnation temperature	78 - 340 K
Running time	several hours
Max. Reynolds number/m	400 million
Drive motor	2.25 MW
Fan speed	up to 6500 rpm
LN_2 tank volume	212 m^3

Table 3   Langley 0.3-m Transonic Cryogenic Tunnel (TCT)

In 1975 we installed a conventional 20 x 60 cm slotted-wall 2-D test section. In 1978 we increased the operating pressure of the tunnel from five to six bars. We used the 20 x 60 cm test section for several airfoil studies as well as studies aimed at developing test techniques for cryogenic tunnels.

In 1986 we installed a 33 x 33 cm solid adaptive wall test section. The test section has single curvature top and bottom walls. The solid sidewalls have provision for boundary layer removal. This test section allows us to test airfoils through transonic speeds at flight values of Reynolds number. We use the adaptive wall test section for both 2-D and 3-D testing. Table 3 gives the major design and operational characteristics of the 0.3-m TCT. Later in this paper I will describe briefly some of the work done in the 0.3-m TCT in its first 16 years of operation.

The Biggest. Along with the early low-speed experimental work on cryogenic tunnels at NASA Langley, there was a parallel theoretical study. Adcock studied in detail the so-called "real-gas effects" of transonic tunnels with nitrogen gas at cryogenic temperatures.[5]   We also had experimental proof of the validity of the cryogenic wind tunnel concept at transonic speeds from the tests we had made in the 0.3-m TCT.

*A single transonic wind tunnel to be called the U.S. National Transonic Facility	
Cryogenic concept	
* Characteristics:	
Test section size	2.5 m x 2.5 m
Design pressure	8.8 atm
Design Mach number	0.2 - 1.2
Stream fluid	Nitrogen
Basic drive power	90 megawatts
Productivity/efficiency	8000 polars/year
Reynolds number	120 million (at M = 1.0)
Located at the NASA Langley Research Center	

Table 4   U.S. National Transonic Facility (NTF) AACB Recommendations

This experimental proof, combined with Adcock's theoretical studies, had far-reaching effects. Researchers abandoned several proposed ambient temperature high Reynolds number intermittent tunnels as the superiority of the cryogenic concept became more widely recognized.

A direct outcome of our work was a 1975 decision by the joint Air Force/NASA Aeronautics and Astronautics Coordinating Board (AACB). The AACB recommended a **single** large transonic **cryogenic** tunnel to meet the high Reynolds number testing needs of the United States. Table 4 shows the AACB recommendations for the proposed **U.S. National Transonic Facility (NTF).**

The NTF at NASA Langley is the manifestation of the AACB recommendations. The site dedication ceremonies for the U.S. NTF took place on July 19, 1977. The dedication of the completed NTF took place on December 6, 1983. The NTF has brought together for the first time the technology of large transonic wind tunnels and modern cryogenic engineering.

There are many interesting details of the NTF design I could describe. However, most of the information that might be of interest is available in references 6 and 7. Table 5 gives the main design and operational characteristics of the U.S. National Transonic Facility.

HEAT TRANSFER TUNNEL

Of special interest because of its unusual use of cryogenic temperatures is the Cryogenic Facility at the University of Illinois at Urbana-Champaign (UIUC). Clausing and his colleagues built a low-speed fan-driven cryogenic tunnel. They have used it for studies of forced, natural, and combined convective heat transfer. It excels in these studies under conditions requiring very large values of both Reynolds number and Grashof number.

Type	closed circuit, fan
Material of construction	304 stainless, aluminum
Insulation	internal
Cooling:	
Cryogenic mode	liquid nitrogen
Air mode	air/water heat exchanger
Test gas	nitrogen or air
Test section size (h,w,l)	2.5 x 2.5 x 7.62 m
Mach range	0.2 - 1.2
Contraction ratio	15:1
Stagnation pressure	1 - 8.9 bars
Stagnation temperature	78 - 340 K
Running time	several hours
Max. Reynolds number/m	480 million
Drive motor	94 MW
Fan speed	up to 600 rpm
LN_2 tank volume	946 m^3

Table 5 U.S. National Transonic Facility (NTF)

The need to predict accurately the combined convective losses from large high temperature solar "power tower" receivers prompted the building of the cryogenic tunnel at UIUC. Clausing saw the cryogenic tunnel as the only way to get the required large values of Grashof and Reynolds numbers with the appropriate and near constant Prandtl number.[8]

Figure 2 shows the variation of Grashof number and Reynolds number with temperature. As noted in reference 8, and as you can see in Figure 2, the use of cryogenic temperatures is a good way to get higher Reynolds number but an even better way to get higher Grashof numbers. Furthermore, the cryogenic environment virtually eliminates the influence of radiative heat transfer which is a major source of error in natural convection data obtained in conventional facilities.[9]

Clausing has reported both the theory and advantages of the cryogenic heat transfer tunnel in more detail in references 9 and 10. The sketch in Figure 3 shows a cross-sectional view of the UIUC Cryogenic Facility.

They cool the tunnel by passing liquid nitrogen through a heat exchanger/vaporizer located just downstream of twin drive fans. The resultant gas vents into the tunnel circuit. In this way they avoid any problems that might arise from incompletely evaporated liquid nitrogen from direct injection. They maintain a slight overpressure during operation to keep out the room air.

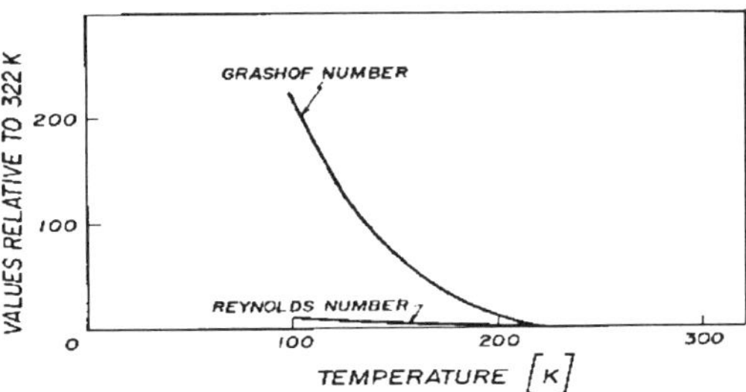

Fig. 2- Effect of Temperature on Grashof and Reynolds number.

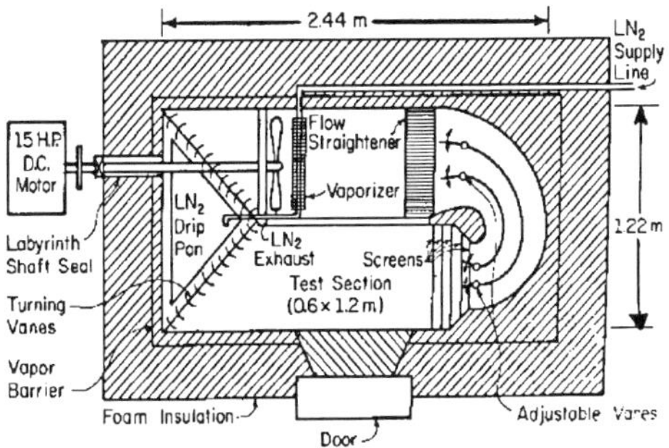

Fig. 3- UIUC Cryogenic Heat Transfer Tunnel

Reference 11 gives a complete description of the UIUC Cryogenic Facility. Table 6 gives the basic specifications of the tunnel.

PLANNED TUNNELS

European Transonic Windtunnel (ETW). Originally working through the Fluid Dynamics Panel of AGARD, four European countries have joined to design and build a large fan-driven transonic cryogenic tunnel in Europe. The tunnel is the European Transonic Windtunnel (ETW). The countries funding the ETW are France, the Federal Republic of Germany, the Netherlands, and the United Kingdom. They expect the ETW to meet their high Reynolds number testing needs at transonic speeds.

Work on the project to date includes building a 1:8.8 scale pilot tunnel at the National Aerospace Laboratory in Amsterdam. The pilot cryogenic tunnel, known as PETW, has the same operating range as anticipated for the ETW. The pre-design phase of ETW has been completed. Each of the four countries is doing research in support of the final design of the ETW. The four countries approved

Type	closed circuit, fan
Material of construction	mostly aluminum
Insulation	external, urethane
Cooling	LN_2 heat exchanger with GN_2 injection
Test gas	nitrogen air for T > 290 K
Test section size (h,w,l), m	1.22 x 0.60 x 1.0
Speed range	0 - 8m/s
Contraction ratio	1:1
Stagnation pressure	atmospheric
Stagnation temperature	80 - 300 K
Running time	several minutes
Max. Reynolds number/m	4 million
Drive motor	11.2 kW
Fan speed	0 - 1750 rpm

Table 6   UIUC Cryogenic Heat Transfer Tunnel

Type	closed circuit, fan
Material of construction	stainless steel
Insulation	internal
Cooling	liquid nitrogen
Test gas	nitrogen
Test section size (h,w,l)	2.0 x 2.4 x 6.9 m
Mach range	0.15 - 1.3
Contraction ratio	12:1
Stagnation pressure	1.25 - 4.5 bars
Stagnation temperature	90 - 313 K
Running time	continuous
Max. Reynolds number/m	228 million
Drive motor	50 MW
Fan speed	up to 1200 rpm
LN_2 tank volume	5000 m³

Table 7   European Transonic Windtunnel (ETW)

the construction phase and construction of the ETW has started on a site adjoining the DLR Center in Koeln, FRG. Reference 12 gives a complete account of the evolution and early development of the ETW. Table 7 gives the major design and operational characteristics of the ETW.

Japan. There are many good wind tunnels in Japan. Some of the best are at Japan's National Aerospace Laboratory (NAL) in Chofu, Tokyo. The wind tunnels at Chofu include the 2 m x 2 m Transonic Wind Tunnel and the 5.5 m x 6.0 m Low-speed Wind Tunnel. These are respectively Japan's largest transonic and subsonic tunnels. However, these tunnels cannot achieve the test Reynolds number needed to develop modern aircraft.

To overcome the problem of low Reynolds number, three groups in Japan developed and are now using cryogenic wind tunnels. There are four cryogenic wind tunnels in Japan. Two are low-speed tunnels at the University of Tsukuba.[13] Another is a 2-D transonic tunnel at the National Defense Academy in Yokosuka.[14]

The fourth cryogenic tunnel in Japan is a transonic tunnel used regularly by Sawada and his colleagues at the National Aerospace Laboratory (NAL).   The NAL 10 x 10 cm Pilot Transonic Cryogenic Tunnel has logged about 600 hours of testing and development work since first running in 1983. References 15 through 19 describe the tunnel and give some of the aerodynamic and operational results.

There are many possible combinations of size and pressure for a transonic cryogenic tunnel to meet the high Reynolds number testing needs of Japan. Sawada has suggested one possible scenario for meeting this need.

Based on his experience with the 10 x 10 cm tunnel, Sawada has suggested two new transonic cryogenic tunnels for Japan. The smaller tunnel would have a 0.6 x 0.6 m test section. This tunnel would obviously be useful in its own right for aerodynamic testing. However, its main purpose would be as a pilot for a second transonic cryogenic tunnel having a 3.0 x 3.0 m test section.

The officials at NAL have not officially endorsed Sawada's suggested cryogenic tunnels. However, his suggestion represents a reasoned approach to providing Japan with a world class high Reynolds number transonic tunnel. Tables 8 and 9 give some of the proposed design characteristics of the two cryogenic tunnels suggested by Sawada.

<u>U.S.S.R.</u> One of my colleagues, Dr. Stephen Wolf, attended a symposium on experimental techniques in Novosibirsk, U.S.S.R. in July of 1988. Dr. Wolf met researchers working on cryogenic tunnels, adaptive wall test sections, and magnetic suspension and balance systems. His trip provided a gold mine of information which he kindly passed on to me.

There are at least three cryogenic wind tunnels operating in the U.S.S.R. Researchers at the Central Aero-Hydrodynamics Institute (TsAGI), Zhukovsky, near Moscow, have built a 0.2 x 0.2 m transonic cryogenic transonic tunnel driven by induction.

At the Institute of Theoretical and Applied Mechanics (ITAM), Novosibirsk, Siberia, researchers have built a low-speed atmospheric tunnel. This tunnel, known as MT-324K, also has a 0.2 x 0.2 m test section.

Researchers at the Physical-Mechanical Institute (PMI-K), Kharjkov, Ukrain, have also built a low-speed cryogenic tunnel which has a 0.22 m circular test section. The purpose of this tunnel at PMI-K is to determine the condensation characteristics of various gases and gas mixtures on cold surfaces. I give more details about the characteristics of these tunnels in reference 20.

Type	closed circuit, fan
Cooling	liquid nitrogen
Test gas	nitrogen
Test section size (h,w,l)	0.6 x 0.6 x 1.8 m
Mach range	0.2 - 1.2
Contraction ratio	14:1
Stagnation pressure	1.2 - 5 bars
Stagnation temperature	90 - 300 K
Running time	45 minutes
Max. Reynolds number/m	340 million
Drive motor	5 MW

Table 8   Proposed NAL Pilot 0.6 m Transonic Cryogenic Tunnel

Type	closed circuit, fan
Cooling	liquid nitrogen
Test gas	nitrogen
Test section size (h,w,l)	3.0 x 3.0 x 6.0 m
Mach range	0.2 - 1.2
Contraction ratio	14:1
Stagnation pressure	1.2 - 9 bars
Stagnation temperature	90 - 300 K
Running time	60 minutes
Max. Reynolds number/m	540 million
Drive motor	90 MW

Table 9   Proposed NAL 3.0 m Transonic Cryogenic Tunnel

Type	closed circuit, drive unknown
Material of construction	unknown
Insulation	unknown
Cooling	liquid nitrogen
Test gas	nitrogen
Test section size (h,w,l)	0.6 x 0.6 x ?? m
Mach range	transonic
Contraction ratio	unknown
Stagnation pressure	1 - 10 bars
Stagnation temperature	80 - 320 K
Running time	unknown
Max. Reynolds number/m	up to about 350 million
LN_2 tank volume	unknown
Drive motor	unknown

Table 10   T-312K Transonic Cryogenic Tunnel

There are two transonic cryogenic tunnels proposed for the U.S.S.R. The first is to be built at Novosibirsk. The ground breaking for this tunnel was scheduled to begin in the autumn of 1988. Table 20 gives the characteristics of this tunnel, to be known as T-312K.

We know even less detail about the proposed U.S.S.R. equivalent to the U.S. NTF. The dimensions are similar to the NTF. The U.S.S.R. NTF is likely to be built alongside other industrial type tunnels at TsAGI Zhukovsky. The design phase of this tunnel was nearly complete in July of 1988.

THE 0.3-M TCT - THE FIRST SIXTEEN YEARS

Earlier I gave some of the design and operational characteristics of the 0.3-m TCT.   In this section I describe some of the work done in the 0.3-m TCT during its first 16 years of operation.

We can arbitrarily divide the work in the 0.3-m TCT into three broad categories.   The first category includes studies to improve the efficiency, safety, and performance of cryogenic tunnels. The second category is the development of test techniques for cryogenic tunnels.   The third category is aerodynamic research.

The majority of the work done in the 0.3-m TCT is reported in the open literature. Rather than cite individual references to the work in this paper, I refer the reader to reference 21, a fairly recent bibliography on cryogenic tunnels. The subject index of reference 21 gives ready access to published papers giving details of work listed in this section.

EFFICIENCY, SAFETY, AND PERFORMANCE

Most cryogenic tunnels use large quantities of liquid nitrogen for cooling. Therefore, cryogenic tunnels are more expensive to operate than the same size and pressure temperature tunnel. This is the price we must pay for high Reynolds number testing.

We must make our high Reynolds number cryogenic tunnels as efficient as possible if we are to take full advantage of their research capability. During the past 15 years we have worked to make the 0.3-m TCT an efficient wind tunnel. Most of the lessons we have learned in the 0.3-m TCT are usable in other cryogenic tunnels.

Safety is an important consideration, especially with pressurized cryogenic tunnels. Again, in using the 0.3-m TCT for over 8000 hours of testing, we have tried to develop safe operating procedures. Table 10 lists some of the work done with the 0.3-m TCT related to efficiency, safety, and performance.

Automatic Controls
- Tunnel conditions
+Mach number
+Pressure
+Temperature
- Adaptive-wall test section
- Side-wall boundary layer removal
GN_2 exhaust
LN_2 injection
- Location
- Nozzle design
LN_2 storage & supply
Operating procedures
- Optimum purging and cooldown
- Minimum energy test direction
Performance
- Condensation boundaries
- Digital valves
Safety
- Organization
- Procedures
- Training
Thermal insulation

Table 11   Work in the 0.3-m TCT related to Efficiency, Safety, and Performance

A contract study of ways to recover energy from the cold tunnel exhaust does not show up in the table.   However, if preliminary studies[22] warrant, we will connect the 0.3-m TCT exhaust to an especially designed expansion engine. Depending on test conditions, this expansion engine would deliver LN_2 to be re-used in the tunnel. We expect recovery rates to be about 10 percent of the exhaust mass flow. This recovery scheme should let us recover some of the energy presently wasted in the cold tunnel exhaust.

Good automatic controls contribute directly to tunnel efficiency, safety, and performance. Being able to hold test conditions constant also improves the quality of the aerodynamic data. From our first day of operation, we have worked toward fully automatic controls for the 0.3-m TCT.

We have recently installed our third generation automatic control system at the 0.3-m TCT. Balakrishna and A. Kilgore have made substantial improvements in tunnel control using a system based on a dedicated minicomputer.[23] The successful development of fully automatic controls for the 0.3-m TCT is one of our major accomplishments.

TEST TECHNIQUES

The minimum goal of any cryogenic tunnel is to be able to do the same types of tests offered by a comparable ambient temperature tunnel. In addition, the cryogenic tunnel should offer some unique testing capability made possible by having temperature as an independent variable.

In the early days of operation, almost every new test in the 0.3-m TCT forced us to work on test techniques. We could use some of our ambient tunnel techniques directly without modification. However, many of them needed modification before we could use them successfully at cryogenic temperatures.

Table 12 lists in rather arbitrary order the major types of test techniques studied in the 0.3-m TCT. Its main purpose is to show the wide range of techniques studied during the past 16 years in the 0.3-m TCT.

Not all of the work listed in Table 12 was successful. For example, we have used several methods of seeding the flow for laser work. One method that worked was a small inadvertent oil leak into the tunnel. Frozen oil does a good job of seeding the flow. However, it also eroded the models. For a variety of reasons, the other methods we tried were not totally satisfactory.[24]

We are still working toward developing test techniques for cryogenic tunnels. As noted in a recent paper by Wolf,[25] much of our effort at the 0.3-m TCT now goes toward developing and using our adaptive wall test section. Work also continues on flow visualization techniques with emphasis on transition detection. Recently, one of our colleagues from NAL completed a year at Langley studying how to do propulsion simulation in cryogenic tunnels.[26]

Building models for cryogenic tunnels is another important area. Without good models it is impossible to take advantage of the testing capability of cryogenic tunnels. Fortunately, we have had considerable success in this area.[27]

Adaptive-wall test section
  - 2-D and 3-D models
2-D Testing
  - Surface pressures (static and dynamic)
  - Wake surveys
  - Oscillating airfoil
  - Non-adiabatic airfoil (LN_2 cooled)
  - Side-wall boundary layer removal
3-D Testing
  - Surface pressures
  - Internal strain-gage balances
Flow angularity probe
Flow visualization
  - Holographic interferometry
  - Shadowgraph
  - Vapor-screen technique
Laser techniques
  - LDV
  - Seeding schemes
  - Two-spot
Model construction techniques
Orifice size effect
Skin-friction measurement
Temperature measurements
Transition detection
  - Fluctuating pressure measurement
  - Hot film gages

Table 12   Test Techniques Studied in the 0.3-m TCT

AERODYNAMIC RESEARCH

For aerodynamic research, the most significant capability of the 0.3-m TCT is the high unit Reynolds number. We can test at unit Reynolds number up to 400 million per meter.

Another extremely important testing capability of the 0.3-m TCT is the very wide range of Reynolds number. This wide range in Reynolds number is a direct result of the wide ranges of operating temperature and pressure.

In incompressible flow (below about M = 0.4) we can cover a 500 to 1 range of Reynolds number. In compressible flow, where Mach number is not an independent variable, we can cover at least a 30 to 1 range.

The range of aerodynamic research in the 0.3-m TCT is fairly broad. Proving the validity of the cryogenic wind tunnel concept at transonic speeds was the aim of the earliest aerodynamic research.

Most of the recent research has been on 2-D airfoils.

Most of the aerodynamic research done in the 0.3-m TCT is reported in the literature.[21] Therefore, I will not go into the details of the work in this paper. However, I will list some of the work to illustrate the research capabilities of the 0.3-m TCT. Table 13 lists the major types of aerodynamic tests made in the 0.3-m TCT.

ADVANCED TESTING TECHNIQUES

Cryogenic tunnels let us test at full-scale Reynolds number. However, full-scale Reynolds number is a *necessary* but not a *sufficient condition* to insure usable data from our wind tunnels. Wall interference and support interference effects can make the data useless by completely masking any Reynolds number effect.

Also, some of our conventional testing techniques introduce problems of their own. One example from the 0.3-m TCT is the wake survey rake we use in airfoil testing. The survey rake becomes a lifting surface in the presence of the downflow from a lifting airfoil. This can cause serious interference effects.

Fortunately, advanced testing techniques exist to solve most of our serious problems. We need to take advantage of these advanced testing techniques to increase the value of cryogenic tunnels to the research community.

ADAPTIVE WALL TEST SECTIONS

To reduce or eliminate wall interference we need to use adaptive wall test sections. Although we include adaptive walls under advanced techniques, British researchers first used them over 50 years ago. Adaptive walls address the problem of wall interference at its source, the test section walls. We can use analytical techniques to correct any wall induced errors left after wall streamlining.

For the 0.3-m TCT we have chosen an adaptive wall test section with solid but flexible top and bottom walls. We have had success with both 2-D and 3-D models through the transonic speed range.[27,28] We and other researchers have demonstrated the practicality of adaptive wall test sections for transonic testing.

MAGNETIC SUSPENSION AND BALANCE SYSTEMS

We need to use magnetic suspension and balance systems (MSBSs) to eliminate completely support interference. The French at ONERA first used magnetic suspension for wind tunnel tests in the mid 1950s. The early success at ONERA led several other researchers to build small systems, mostly for hypersonic research. In 1979, Britcher demonstrated the combination of a 6- component MSBS with the low-speed cryogenic tunnel at the University of Southampton.[29] Britcher took data on a body of revolution with tunnel temperatures below 100 K.

Advances in technology make it possible to build large MSBSs for large wind tunnels. Recent design studies confirm the feasibility of building systems for 2 or 3 m test sections. Further, testing experience at several laboratories has demonstrated the research potential of MSBS.

2-D Testing
- Effect of Reynolds number on performance
- Circular cylinder
- Non-adiabatic wall effects
- Surface roughness effects

Semi-span Testing
- Buffet
- NACA 64A010 airfoil (flutter)
- X29A canard

3-D Testing
- Shuttle Orbiter
- Boattail models
- Wing-body interference

Special Testing
- NTF cooling coil
- Tunnel wall boundary layer

Table 13   Types of Aerodynamic Research in the 0.3-m TCT

LASER TECHNIQUES

The use of laser techniques to eliminate intrusive measurement methods is a logical and essential step if we are to take full advantage of our new found high Reynolds number capability of cryogenic tunnels

Taken together, we have solutions to most of the problems with subsonic and transonic wind tunnels. As we continue to develop and apply these solutions, we will enter a new era of experimental aerodynamics. Improved experimental data will lead inevitably to improvements in computational fluid dynamics techniques. Likewise, improved computational techniques will lead to better experimental data. This symbiotic relationship will help aerodynamics move with confidence into the 21st century.

CONCLUSIONS

1. Cryogenic tunnels are in regular and productive use around the world.

2. The first sixteen years of operation of the 0.3-m TCT has been safe, productive, and a cryogenic engineering success.

3. We must use advanced testing techniques, such as adaptive walls, MSBS, and laser techniques, to take full advantage of cryogenic tunnels.

4. The use of cryogenic tunnels will increase dramatically as we begin to use high Reynolds number data to develop and verify advanced CFD codes.

5. Cryogenic wind tunnels will join with supercomputers as our basic tools as we move the aeronautical sciences into the 21st century.

REFERENCES

1. D.A. Dress and R.A. Kilgore, **Cryogenic Wind Tunnels - A Global Perspective.** Cryogenics, vol. 28, January 1988, pp. 10-21.

2. R.A. Kilgore, M.J. Goodyer, J.B. Adcock, and E.E. Davenport, **The Cryogenic Wind Tunnel Concept for High Reynolds Number Testing.** NASA TN-D-7762, November 1974, 96 pp.

3. G. Viehweger, **The Kryo-Kanal-Koeln Project, KKK.** Paper no. 4 in AGARD-R-774, 1989.

4. R.A. Kilgore, J.B. Adcock, and E.J. Ray, **Simulation of Flight Test Conditions in the Langley Pilot Transonic Cryogenic Tunnel.** NASA TN-7811, December 1974, 24 pp.

5. J.B. Adcock, **Real-Gas Effects Associated With One-Dimensional Transonic Flow of Cryogenic Nitrogen.** NASA TN-D-8274, December 1976, 272 pp.

6. Anon: **Cryogenic Technology.** NASA CP-2122, Parts I and II, March 1980, 441 pp.

7. W.E. Bruce Jr., **The U.S. National Transonic Facility, NTF.** Paper no. 3 in AGARD-R-774, 1989.

8. A.M. Clausing, G.L. Clark, and M.H. Mueller, **The Cryogenic Heat Transfer Tunnel - A New Tool for Convective Research.** Presented at the Winter Annual Meeting, ASME, San Francisco, California, 1978, pp. 73-78.

9. A.M. Clausing, **Experimental Studies of Forced, Natural and Combined Convective Heat Transfer at Cryogenic Temperatures.** Paper 24, 1st International Symposium on Cryogenic Wind Tunnels, Southampton, U.K., 1979, 8 pp.

10. A.M. Clausing, **Advantages of a Cryogenic Environment for Experimental Investigations of Convective Heat Transfer.** International Journal of Heat and Mass Transfer, vol. 25, no. 8, 1982, pp. 1255-1257.

11. M.H. Mueller, et al: **Description of UIUC Cryogenic Wind Tunnel Including Pressure Distributions, Turbulence Measurements and Heat Transfer Data.** University of Illinois Technical Report ME-TN-79-9180- 1, 1979, 82 pp.

12. X. Bouis, **The European Transonic Windtunnel, ETW.** Paper no. 6 in AGARD R-774, 1989.

13. T. Adachi, **Cryogenic Wind Tunnel and its Activities in University of Tsukuba.** June 19, 1987, Personal Communication.

14. Y. Yamaguchi, N. Kuribayashi, and H. Kaba, **Characteristics for ambient condition of NDA Cryo-tunnel and an attempt on its automatic cryogenic operation.** Proceedings of the 19th Annual Meeting of JSASS, April 5-6, 1988, pp. 73-74.

15. K. Takashima, H. Sawada, T. Aoki, and S. Kayaba, **Trial Manufacture of NAL 0.1m x 0.1m Transonic Cryogenic Wind Tunnel.** NAL TR-910, 1986, pp 58.

16. H. Sawada, **NAL TCWT Status - Cryogenic Operation.** NAL News, 1984-3, No. 229.

17. H. Sawada, **Automatic Operation of the NAL Cryogenic Wind Tunnel.** NAL News, 1986-1, No. 321, pp. 2-4.

18. H. Sawada, **Heated External Balance for Cryogenic Wind Tunnel.** NAL News, 1987-1, No.. 333, pp. 2-3.

19. H. Sawada, **Cryogenic Wind Tunnels.** Journal of JSASS, June 1987, pp. 285-293.

20. R.A. Kilgore, **Other Cryogenic Wind Tunnel Projects.** Paper no. 9 in AGARD R-774, 1989.

21. M.H. Tuttle, R.A. Kilgore, and K.L. Cole, **Cryogenic Wind Tunnels - A Selected, Annotated Bibliography.** NASA TM-4013, September 1987, 89 pp.

22. G.E. McIntosh, D.S. Lombard, D.L. Martindale, and R.P. Dunn, **Cost Effective Use of Liquid Nitrogen in Cryogenic Wind Tunnels--Final Report.** NASA CR-178279, April 1987, 62 pp.

23. S. Balikrishna, and W.A. Kilgore, **Minicomputer Based Controller for the Langley 0.3-m Transonic Cryogenic Tunnel.** NASA CR-181808, March 1989, 148 pp.

24. P.L. Lawing, and C.B. Johnson, **Summary of Test Techniques Used in the NASA-Langley 0.3-m Transonic Cryogenic Tunnel.** AIAA Paper 86-0745, March 1986, 11 pp.

25. S.W.D. Wolf, and E.J. Ray, **Highlights of Experience with Flexible Walled Test Section in the NASA Langley 0.3-meter Transonic Cryogenic Tunnel.** AIAA Paper 88-2036, May 1988.

26. K. Asai, **Hot-Jet Simulation in Cryogenic Wind Tunnels.** NASA RP-1220, July 1989, 46 pp.

27. P.L. Lawing, **The Construction of Airfoil Pressure Models by the Bonded Plate Method: Achievements, Current Research, Technology Development and Potential Applications.** NASA TM-87613, Sept. 1985, 31 pp.

28. C.L. Ladson, and E.J. Ray, **Status of Advanced Airfoil Tests in the Langley 0.3-Meter Transonic Cryogenic Tunnel.** Paper in NASA CP-2208, September 14-15, 1981, pp. 37-53.

29. R. Rebstock, and E.E. Lee Jr., **Capabilities of Wind Tunnels with Two Adaptive Walls to Minimize Boundary Interference in 3-D Model Testing.** NASA CP - 3020, vol. I, Part 2, March 1989, pp. 891-910.

30. C.P. Britcher, and M.J. Goodyer, **The Southampton University Magnetic Suspension/ Cryogenic Wind Tunnel Facility.** Paper 10, 1st International Symposium on Cryogenic Wind Tunnels, Southampton, U.K., 1979, 9 pp.

AERODYNAMIC TESTING IN CRYOGENIC NITROGEN GAS - A

PRECURSOR TO TESTING IN SUPERFLUID HELIUM

Pierce L. Lawing

NASA, Langley Research Center
Hampton, VA 23665-5225
U.S.A.

SUMMARY

This paper brings together some of the testing techniques developed for transonic cryogenic tunnels using Nitrogen as the test fluid. It emphasizes techniques which required special development, or were unique because of the opportunities offered by cryogenic operation. The first part of the paper is used to discuss measuring the static aerodynamic coefficients normally used to determine component efficiency. The first topic is testing of two dimensional airfoils at transonic Mach numbers and flight values of Reynolds number. Three dimensional tests of complete configurations and sidewall mounted wings are also described. Since flight Reynolds numbers are of interest, free transition must be allowed. A discussion is given of wind tunnel and model construction effects on transition location.

The second part of the paper deals with time dependent phenomena, fluid mechanics, and measurement techniques. The time dependent, or unsteady, aerodynamic test techniques described include testing for flutter, buffet, and oscillating airfoil characteristics. In describing non-intrusive laser techniques, discussions are given regarding optical access, seeding, forward scatter lasers, two-spot lasers, and laser holography. Methods of detecting transition and separation will be reported and a new type of skin friction balance is described. There are 29 references to assist in obtaining further information.

INTRODUCTION

The advantages of conventional cryogenic for wind tunnel testing are, by now, well documented and understood, references 1-3. However, the development of cryogenic testing tools is an ongoing process. Also, in order to properly take advantage of all of the research opportunities offered by cryogenic operation of a wind tunnel, it is necessary to develop some new techniques unique to cryogenic operation. In addition, the fact that the cryogenic facility can produce realistic flight boundary layers requires a quantum improvement in model construction techniques, involving surface finish, fidelity of model contours, and dimensional stability. A similar effort is required in instrumentation technique, particularly instrumentation techniques for transition detection and the study of unsteady aerodynamics.

It is now necessary to reconsider our testing techniques and opportunities with the new concept of testing in superfluid helium. This paper will rely heavily on the development of testing techniques in the Langley 0.3-m Transonic Cryogenic Tunnel, or 0.3-m TCT, and will summarize techniques used which either had special development required by the cryogenic environment, or are unique because of the opportunities offered by cryogenic operation. There are about 20 techniques listed and the author has been directly involved in the majority of these experimental efforts. Several of the techniques for which there is little or no information elsewhere are discussed in some detail.

AIRFOIL TESTING

Two-dimensional airfoil testing is a standard test technique used to examine the characteristics of an airfoil section in more detail than would be possible in a full aircraft configuration. One problem peculiar to this type of testing is error introduced by the penetration of the model through the thick, tunnel-side-wall, boundary layer. Since this boundary layer does not have as much momentum as the external stream, it will separate much easier than the free stream flow. At low lift conditions, separation may not occur. As the lift increases, the side wall boundary layer will separate. At high lift, this separated region becomes large enough to influence measurements at the center of the airfoil, thus spoiling the data.

Other errors are introduced by the presence of the floor and ceiling of the test section. One serious error is caused by the blockage of the model, especially at transonic conditions. This is commonly alleviated by slots in the walls, or in the current 0.3-m TCT test section, by adaptive walls. Methods of dealing with these error sources are described in the next sections.

Airfoil Test With Sidewall Boundary Layer Removal

Figure 1 shows pressure surveys through the wake of a 12 percent thick supercritical airfoil at three spanwise stations. Results are shown with and without "bleed". The term here refers to the removal of the tunnel sidewall boundary layer just ahead of the model to remove some of the low energy boundary layer fluid. This action has the effect of increasing the boundary layer's resistance to separation. In addition, the sidewall boundary layer interference is reduced because of the reduction in the displacement thickness, as indicated by the values of the parameter $2\,\delta^*/b$, shown on figure 1. The case presented is for a high value of lift coefficient and at a transonic flow condition. The Gaussian shaped part of the curve is due to the viscous drag losses and the ragged ramp portion is due to the shock losses. The use of boundary layer removal smoothes the wake distributions, especially the shock loss part of the curve, as well as making the shape of the curves more uniform in the spanwise direction. For additional details and more information see references 4 and 5.

Mass flow may be removed from the boundary layer either by passive bleed to the atmosphere or by a compressor. The compressor, shown schematically in figure 2, must operate over a wide range of pressure and temperature (ambient to near liquid nitrogen). The mass removed from the sidewalls of the test section is reinjected. During compressor validation tests, reference 6, with sidewall boundary layer removal and reinjection, tunnel flow conditions could easily be maintained over a wide range of Mach numbers and Reynolds numbers. The successful integration of the compressor with the normal tunnel operation was not only a significant engineering achievement in its own right, but the use of the compressor greatly enhances the tunnel's research capabilities. An important aspect for two dimensional airfoil research, is the capability to treat the test section sidewall boundary layers to reduce sidewall interference. Recent improvements to the compressor/tunnel control system are given in reference 7.

Adaptive Wall Test Section

To reduce or eliminate wall interference we need to use **adaptive wall test sections**. Although we include adaptive walls under advanced techniques, British researchers first used them 50 years ago, reference 8. Modern digital controls and powerful computers have made the application of adaptive wall technology an easier process. Adaptive walls address the problem of wall interference at its source, the test section walls. We can use analytical techniques to correct any wall induced errors left after wall streamlining. For the 0.3-m TCT we have chosen an adaptive wall test section with solid but flexible top and bottom walls. We have had success with both 2-D and 3-D models through the transonic speed range. We and other researchers have demonstrated the practicality of adaptive wall test sections for transonic testing.

NEW TECHNIQUE FOR AIRFOIL PRESSURE MODELS

Building models for the new class of cryogenic tunnels requires advanced new technologies. In order to isolate the sought after Reynolds number effects from surface roughness induced effects, it is necessary to have very good surface finish (6 to 8 microinches at the leading edge). Special materials are utilized to provide the necessary fracture toughness at cryogenic temperatures. Exceptionally good dimensional stability is required during temperature cycling to maintain the accuracy demanded of transonic models. These requirements, together with the required high strength, make the task of building the necessary models difficult, leading to lost research time and increased model costs.

In the early years of 2-D testing in the 0.3-m TCT, the difficult fabrication task led to a rejection rate of more than 50% of the models attempted, and several parallel efforts to find new methods of building the necessary models were initiated. Improved application of the more or less conventional model building activities was one such effort, and it is documented in reference 9. Other efforts include adhesively bonded cover plates reference 10, steel pressure tubes cast in a fiberglass airfoil, and the bonded plates method described below.

The technique pictured in figure 3 is a promising new solution being researched to alleviate the model construction problem for pressure instrumented models. Here, the pressure channels are cut or etched into the opposing faces of a metal "sandwich"; the sandwich is closed and brazed together in a vacuum oven resulting in a model "blank" with high strength and low "plumbing" costs. This model has been built, successfully tested, and the data reported in reference 11. Sample test results are shown in figure 4. Part (a) of figure 4 is for the model at an angle of attack of 2 degrees, corresponding to a lift coefficient of 0.31. Part (b) is the data of part (a) with the data for -2 degrees, lift coefficient of -0.31, superimposed. Figure 4 presents pressure coefficient data from the model described above, at transonic conditions and a moderate lift coefficient. This data shows the presence of a shock wave on the upper surface between 30 and 38 percent chord. The circular symbols are the upper surface pressure taps and the square symbols are for the lower surface. Since this is a symmetrical airfoil, the pressure signatures at positive and negative angle of attack should match with the upper and lower symbols changing places. As may be observed from part b, the pressures are very nearly the same. This demonstrates good top and bottom symmetry of model construction and good testing technique. For information on present research for more advanced models including electrical discharge wire cut of airfoil contours, thin airfoils, etched pressure channels and curved bond planes, see references 12 and 13.

3-D AIRFOIL TESTING

Thin Wing Testing

During our airfoil testing program in the 0.3-m TCT, we were never successful in building a pressure instrumented airfoil any thinner than 10% of chord. Since many of our military aircraft operate with wings much thinner than this, the lack of thin airfoil capability was a serious deficiency. In the course of the model construction research described above it became evident that thin airfoils were now possible. A model of the X29 fighter canard was built; the thickness was 5%. The model was tested primarily to verify the model construction technique, but also the test provided the opportunity to take unique aerodynamic data and to exercise the solid adaptive wall test section, reference 14, for 3-D testing. The airfoil was tested over nearly the full range of conditions available with the adaptive wall test section. This included both ambient and cryogenic operation at pressures up to 6 bars at Mach numbers up to 1.07. For testing this semi-span 3-D model, the

method of Rebstock, reference 15, was used to adapt the solid test section walls for minimum interference.

Aerodynamic data was taken for the airfoil (surface pressures and wake survey) at Mach numbers from 0.3 to 1.07. The tests covered angles of attack from - 4° to + 15°. At most Mach numbers a 10 to 1 range of Reynolds number was covered and flight values of Reynolds number were easily obtained. Figure 5 shows an example of the airfoil data at an angle of attack of 8°, a Mach number of 0.9, and a Reynolds number based on mean aerodynamic chord of 32 million.

Mapping of Three Dimensional Wakes by Pitot Surveys

Figure 6 illustrates a technique of creating a topology of the momentum loss in the wake of a body, in this case a cylinder. Wake surveys with a vertical rake of pitot tubes are a conventional method of determining the momentum loss, and thus the drag of a body. These data were obtained by an updated version of this method where the survey rake is driven through the wake by a computer controlled actuator, and there are six tubes, in a spanwise row, which allow determination of the spanwise variation in the wake. Not only does conducting such a survey provide additional data, but doing so in a cryogenic environment is an engineering achievement. This particular sample illustrates the classic Gaussian shaped curve for those tubes near the tunnel centerline, and shows deviations from this shape at stations close to the wall. This is due to the interaction between the wall boundary layer and the cylinder. Reference 16 includes additional data.

Evaluation of a Cryogenic Strain-gage Balance

Although the 0.3-m TCT has been equipped with a special test section for two dimensional airfoil testing in recent years, it is possible to do small scale three dimensional testing as shown in figure 7. Here an airfoil shape is used to support a sting and a conventional 3-D model arrangement to allow force balance testing. The strain-gage balance is shown between the model and the 2-D support. The balance was tested at cryogenic temperatures with and without electrical resistance heaters, achieving good results in both cases. Further details are contained in reference 17.

UNSTEADY AERODYNAMICS

Flutter Model Testing

A flutter test has been conducted in the 0.3-m Transonic Cryogenic Tunnel to explore problems, develop testing techniques, and determine the potential of a cryogenic tunnel to advance the state of the art in flutter testing. A simple "text book" rectangular planform wing model supported by a beam flexure was used for the test. Model and support were machined from a single piece of 18 Nickel grade 200 maraging steel (trade name Vascomax 200). This material is characterized by its good dimensional stability with temperature change and its high fracture toughness at cryogenic temperatures. Although no "hard" flutter points were included in the test, the model oscillations were large enough to be easily visible on a video monitor at conditions near flutter onset. Figure 8 presents a comparison of analytical and experimental flutter results in terms of the flutter dynamic pressure as a function of Mach number. It is presented here only to illustrate flutter data taken at cryogenic temperatures, and was taken from reference 18. Further details including the effect of Reynolds number on transonic flutter are also contained in this reference.

Buffet Testing Technique

Buffet testing has been conducted in the 0.3-m TCT using semi-span models mounted on one turntable. Instrumentation included a root bending gage to indicate buffet onset. Models included both delta and straight wing planforms. The ability to hold Mach number and Reynolds number constant while varying the velocity demonstrated the strong dependence of buffet onset on the reduced frequency, figure 9. This segment of the research is documented in reference 19. This research is ongoing and recent tests have used carbon composite models to increase the resonant frequency of the model.

Oscillating Airfoil Testing

Figure 10 is a photograph of the 14 percent supercritical airfoil model mounted in a test section module of the 0.3-m TCT. The model is being viewed from the trailing edge. The drive bellows that transmitted the torque to oscillate the model from an external actuator is visible attached to the left wall of the module. The model was oscillated at frequencies as low as 4 cycles per second, and as high as 60 cycles per second. Amplitudes were as low as $\pm$ 1/4 degree

and as high as $\pm$ 1.0 degrees. Tests were conducted at cryogenic temperatures and well into the transonic speed range. Fluctuating pressure data was measured by 43 pressure transducers mounted in the interior of the model. This data will be helpful in the understanding of unsteady aerodynamic processes including flutter, buffet, and rapid changes in pitch over a range of Reynolds numbers including values representative of flight.

NON-INTRUSIVE FLOW MEASUREMENTS

The term non-intrusive implies optical access through the wind tunnel wall, usually a window. For cryogenic facilities operating at high pressures and cryogenic temperatures, such windows pose major problems. For instance, in the 0.3-m TCT, a quartz window shrinks much slower with decreasing temperature than the aluminum structure. At room temperature, the quartz piece must fit loosely, and provide room for the aluminum to shrink around the quartz as the temperature is lowered to cryogenic operation conditions. At the same time, the window and frame must support the pressure drop at room temperatures.

A window at cryogenic temperatures exposed to room temperature air and its attendant humidity must be kept clear of fog and/or frost. One method of maintaining a clear window is to position a thin piece of glass about 10% of the window diameter away from the quartz and purge the resulting gap with dry, heated, Nitrogen gas. Since this is a transonic tunnel with a plenum chamber, there must be a window in the test section wall as well. Thus an optical beam must traverse a thin, unloaded, piece of glass with a small thermal gradient, the purge gas region, the heavily loaded, up to 6 atmospheres, pressure shell window, the nitrogen in the plenum, the lightly loaded, essentially isothermal, test section window, the tunnel wall boundary layer, and finally the test flow. In the case of some techniques, such as a shadowgraph, the process is repeated in reverse on the other side of the test section. In the case of backscatter laser measurements, the low intensity signals must travel back through the same set of conditions.

There are several additional difficulties to recognize in the application of optical techniques to the 0.3-m TCT: Since all of the structural parts must contract with decreasing temperature, and it is not possible to insure isothermal structures, provision must be made for referencing measurements in space to the model locations. For example, this required a special alignment laser for laser velocimeter measurements, reference 20. Other sources of error that must be considered are deflection of structural parts due to differential pressures or model support loads and effects such as lensing of the windows under distortion from pressure or thermal loading. Most of these difficulties are inherent in a facility designed to provide aerodynamic coefficients at flight conditions as opposed to a facility designed for pure fluid mechanics studies or for validation of analytical methods.

Optical Methods

Various flow diagnostic and visualization techniques have been tried in the 0.3-m TCT both to enhance the research utility of the tunnel and to explore the problems and opportunities offered by cryogenic testing. The use of the Laser Transit Anemometer to survey velocity distributions in flow fields and boundary layers is described elsewhere in this paper and in reference 16. Other methods, references 21-25, include laser holographic inteferometry, schlieren, shadowgraph, image quality studies, and moire deflectometry measurements. All have been successful to some extent. One major problem that was encountered was a swamping of the flow features by an optical disturbance that strengthened as the tunnel temperature was lowered. This problem was isolated and identified as thermal inhomogeneities external to the test section, primarily due to convective currents in the plenum chamber. The optical degradation becomes more severe with decreasing temperature and increasing pressure, and exhibits a $(p/T)^2$ dependence. Figure 11 shows a comparison of optical quality before and after isolation of the optical beam from the plenum. This problem is discussed in detail in reference 25.

Seeding a Cryogenic Tunnel

Several of the more promising laser diagnostic techniques require the flow to be seeded with reflective particles a few microns in diameter. At this time there is no clearly satisfactory method of seeding a cryogenic tunnel. Measurements have been made using "natural" seeding generated by running the tunnel cold enough to preserve condensed nitrogen droplets, or by pulsing the liquid nitrogen control to generate a temporary cloud of liquid nitrogen droplets. A chance oil leak past the fan shaft seal provided very satisfactory data rates when the flow was cold enough to promote condensation of the oil into

droplets. However, these schemes have serious drawbacks in that the errors introduced by testing in condensed flow are not defined and no correction method exists, the flow conditions are not known during injection pulsing, and at the extreme cold necessary for high Reynolds number operation, it is suspected that the oil freezes into solid particles and is responsible for eroding the model leading edges. Seeding has also been accomplished by bleeding service air into the tunnel and reducing the temperature to form condensed water, or ice, depending on the temperature. Once again, the larger condensates are expected of causing model erosion at the lower temperatures. To offset the fan heat, the tunnel must be continually injected with liquid nitrogen and the resulting gas vented. Thus the facility continually purges itself, and seeding material must be constantly replenished. Also, since the material is continually vented, it must be environmentally acceptable.

High Reynolds number testing requires good model surface finish, particularly on the leading edges, and small pressure orifices. Great care must be exercised to preserve the finish and keep the orifices unblocked even before the model is installed into the tunnel; introduction of solid particles is advisable only if they do not aggregate and form clumps that have sufficient ballistic coefficient to impact the model, and if they cannot accumulate in the model orifices or other sensitive mechanisms such as actuators. Liquid particles must not form ice clumps at low temperatures and must evaporate at room temperatures to allow cleanup. Thus far the only solid particles that have been tried are kaolin powders. The liquids are water and lubricating oil. Both the kaolin and the oil required extensive cleaning of the tunnel after use and are considered unsuitable materials.

Two-spot Laser Boundary Layer Survey

The two-spot laser, or more properly the Laser Transit Anemometer, LTA, focuses two laser beams into spots 9 microns in diameter with a spot-to-spot separation of about 20 spot diameters. When the spots are aligned with the flow in a wind tunnel, that is with one spot upstream of the other, any reflective particle in the flow that happens to pass through the upstream spot will reflect back a pulse of light. As the particle passes through the second spot it will reflect back a second pulse of light. Since the distance between the spots is known, a measurement of the time between the two pulses yields velocity. This process may be greatly enhanced by seeding the flow with appropriate particles. The data in figure 12 was generated by such a process where the LTA was used to survey the flow field of a cylinder along the line indicated in the "scan location" inset. This data is unique in that it was taken in a cryogenic environment and is one of the first successful attempts at measuring points in the boundary with this type of device. The two decreasing velocity points nearest the cylinder surface are in the boundary layer. Further details are available in reference 16.

FLOW QUALITY

Measurement of Fluctuating Pressures in the Settling Chamber

Fluctuating pressures were measured on the settling chamber sidewall using a commercially available pressure transducer. Figure 13a presents the RMS fluctuating pressure normalized by local mean static pressure as a function of test section Reynolds number for a range of test section Mach numbers. These data cover a large portion of the tunnel operating envelope at a temperature of 140K. Figure 13b presents the spectra of the pressure fluctuations versus the normalized frequency, nondimensionalized by the revolutions of the fan, n, and the number of fan blades, N. The correlation of the data for fundamental frequency and approximately the first eight harmonics clearly indicates that the peaks in the fluctuation pressures are from the fan blade passage. Additional information is available in reference 26.

Measurements of Free Stream Turbulence by 3-wire Hot Wire Probe

The heat transfer characteristics of a three-wire hot-wire probe operated with a constant temperature anemometer were investigated in the subsonic compressible flow regime. The sensitivity coefficients, with respect to velocity, density and total temperature, were measured and the results were used to calculate the velocity, density, and total temperature fluctuations in the test section of the Langley 0.3-m TCT. The test section velocity and total temperature fluctuations are shown in figure 14. In the past mass flow fluctuations, referred to as velocity fluctuations, were often used to determine the disturbance levels in transonic wind tunnels. When the mass flow fluctuations for the 0.3-m TCT were calculated from the correlation coefficient of density and velocity, and the density and velocity fluctuations (see fig. 14a)

from the 3 wire probe data the results were approximately an order of magnitude lower than the results shown. One should be aware of the difference between disturbance level determined from mass flow fluctuations and the 3 wire probe method. The method for the calibration and data reduction of a three wire probe is described in reference 27. By using the three wire probe technique, velocity, density and total temperature fluctuations can be measured without any further assumptions other than those usually made for hot wire anemometry in flows with small perturbations.

Measurement of Flow Dynamics in the Settling Chamber

There is theoretical and experimental evidence which indicates that a sudden or step increase or decrease in the rate of liquid nitrogen injection into the circuit of a cryogenic wind tunnel can cause a temperature front (or step) in the flow for several tunnel circuit times. Since these fronts can have an effect on the control of the tunnel as well as the time required to establish steady flow conditions in the test section of a cryogenic wind tunnel, tests were conducted in the settling chamber in the Langley 0.3-m TCT using high response instrumentation to measure the possible existence of these temperature fronts. Three different techniques were used to suddenly change the rate of liquid nitrogen injection into the tunnel and the results from these three types of tests showed that temperature fronts do not appear to be present. Figure 15 illustrates the "manual step" decrease in total temperature with no indication of temperature fronts for a circuit time of 0.89 seconds. See reference 26 for further details.

FLUID MECHANICS

Skin Friction Measurements

A skin friction balance has been developed by the University of Tennessee Space Institute, with the cooperation and support of the Experimental Techniques Branch, NASA Langley, for test on the test section sidewall of the 0.3-m TCT. The balance uses a belt supported by two flexures to measure shear force. Figure 16 is a section drawing of the device indicating the location of the flexures and the belt. The entire device is mounted with the the belt flush with the surface of interest, in this case the test section side wall. Figure 17 is a plot of skin friction coefficient measured on the test section wall of the 0.3-m TCT and shown as a function of momentum thickness Reynolds number. Data was obtained at temperatures from ambient down to 100 K. The weak change with Reynolds number exhibited, rather than the classic power law decay, is thought to be due roughness due to joints between tunnel sections upstream of the measuring station. Cursory comparison with a Moody pipe flow chart indicates an equivalent sand roughness height of 0.005 inches is sufficient to cause this type of Reynolds number trend. An interesting potential of this balance occurs with the substitution of a fiber optic pickup for the usual strain gages, which removes the barriers to miniaturization of the balance to a size where it can be installed into an airfoil model. These figures and further details may be found in reference 28.

Testing of Controlled Nonadiabatic Airfoils

The coupling of transonic flow and heat transfer has potential benefits both in boundary layer control and to enhance our understanding of fluid mechanics. The case of the model cooler than the flow is of particular interest, but the experiment is difficult in a conventional tunnel due to the formation of frost and ice. In the 0.3-m TCT cryogenic tunnel however, the working fluid is dry nitrogen and there is no water vapor to condense on the model. Testing is conducted by operating the tunnel above the temperature of liquid nitrogen and, by forcing liquid nitrogen through the model using the tubes shown in figure 18, it is possible to maintain the model surface at a much lower temperature. The model is constructed of beryllium-copper, which has a very high thermal conductivity, to aid in maintaining nearly uniform surface temperatures in the presence of heat transfer to the test gas. The model was equipped with sufficient surface pressure orifices to determine chord-wise pressure distributions. The integrated distributions were used together with the wake rake to calculate lift, drag, and pitching moment. These were determined for wall to total temperature ratios of adiabatic, slightly less than adiabatic, and as low as approximately 0.5.

TRANSITION DETECTION

Transition Detection With Specialized Hot Films

An investigation to determine the location of boundary-layer transition was carried out using a similar cooled model to the one described above. The photograph in figure 19 shows the model mounted in the 0.3-m TCT with the chordwise rows of hot-film gages mounted on the upper surface. The model is a 9 inch chord, 12 percent supercritical 2-D airfoil and was instrumented with 48 hot-films. The extremely thin hot-films were applied to the surface of the model by the Douglas Aircraft Company (DAC), using a newly developed method, as part of a NASA/DAC cooperative program to develop a specialized system for detecting boundary-layer transition in cryogenic wind tunnels. The tests, conducted in the Langley 0.3-m TCT, were done both at an adiabatic wall condition and at a non-adiabatic wall condition with liquid nitrogen circulation through the model to cool the surface below the adiabatic recovery temperature. The surface cooling was done to determine the effect of wall temperature on the location of boundary-layer transition at wall to total temperature ratios as low as 0.47. The test results indicated, that with the proper electronic data acquisition equipment, an "on-line" location of boundary-layer transition could be obtained both at ambient and cryogenic conditions. The on-line signal from the hot-films clearly indicated either a laminar, transitional or turbulent boundary layer. Preliminary results indicated that model cooling actually decreased the transition Reynolds number due to the apparent dominance of surface roughness on transition at this condition.

Transition Location by Fluctuating Pressure Measurements

Figure 20 shows the outline of a 14 percent thick airfoil recently tested in the 0.3-m TCT. This two-dimensional model was largely hollow and contained 43 transducers capable of measuring fluctuating pressures. The three traces shown above the airfoil are typical of the output data from these transducers. The trace at the left is near the nose of the model, has a low length Reynolds number, and exhibits a low amplitude fluctuating pressure trace typical of laminar flow. The center trace is from a transducer measuring pressures further back on the airfoil and produces a trace typical of transitional flow with a higher fluctuating amplitude as background and with superimposed turbulent precursor bursts of much higher amplitude. The trace at the right is well into fully developed turbulent flow characterized by high amplitude, high frequency fluctuations in pressure. High amplitude fluctuations in the turbulent regime are perhaps expected for pressure measurements, as opposed to the decreased amplitudes recorded by the heat transfer devices, since the thickening boundary does little to attenuate pressure waves, but generally lowers hear transfer rates. Reference 29 presents data that indicates this same trend.

SUMMATION

Sophisticated test techniques have been developed to take advantage of the many research opportunities offered by the 0.3-m Transonic Cryogenic Tunnel. Although this review has not been exhaustive, sufficient material has been presented to show that test techniques used in the 0.3-m TCT are on par with the most advanced transonic facilities and that cryogenic operation poses no insurmountable handicaps.

REFERENCES

1. Tuttle, Marie H.; Kilgore, Robert A.; and McGuire, Peggy D.: **Cryogenic Wind Tunnels - A Selected, Annotated Bibliography.** NASA TM-86346, April 1985.

2. Kilgore, R. A.; and Dress, D. A.: **The Application of Cryogenics to High Reynolds Number Testing in Wind Tunnels. Part 1: Evolution, Theory, and Advantages.** Cryogenics, Vol. 24, August 1984, pp. 395-402.

3. Kilgore, R. A.; and Dress, D. A.: **The Application of Cryogenics to High Reynolds Number Testing in Wind Tunnels. Part 2: Development and Application of the Cryogenic Wind Tunnel Concept.** Cryogenics, Vol. 24, September 1984, pp. 484-493.

4. Johnson, C. B.; Murthy, A. V.; Ray, E. J.; Lawing, P. L.; and Thibodeaux, J. J.: **Effect of Upstream Sidewall Boundary Layer Removal on an Airfoil Test.** NASA CP 2319, 1983.

5. Murthy, A. V.; Johnson, C. B.; Ray, E. J. and Stanewsky, E.: **Investigation of Sidewall Boundary Layer Removal Effects on Two Different Chord Airfoil Models in the Langley 0.3-Meter Transonic Cryogenic Tunnel.** AIAA Paper No. 84-0598, 1984.

6. Johnson, C. B.; Murthy, A. V.; and Ray, E. J.: **A Description of the Active and Passive Sidewall-Boundary-Layer-Removal Systems of the 0.3-Meter Transonic Cryogenic Tunnel.** NASA TM-87764.

7. Balakrishna, S.; Kilgore, Allen W.; and Murthy, A. V.: **Performance of the Active Sidewall Boundary-Layer Removal System for the Langley 0.3-Meter Transonic Cryogenic Tunnel.** Vigyan Research Associates, Inc.; Hampton, VA 23665-1325. NASA CR 181793, Feb.1989.

8. Wolf, S.W.D.; and Ray, E.J.: **Highlights of Experience with Flexible Walled Test Section in the NASA Langley 0.3-meter Transonic Cryogenic Tunnel.** AIAA Paper 88-2036, May 1988.

9. Young, C. P.; Bradshaw, J. F.; Rush, H. F., Jr.; Wallace, J. W.; and Watkins, V. E., Jr.: **Cryogenic Wind Tunnel Model Technology Development Activities at the NASA Langley Research Center.** Paper No. AIAA 84-0586, 1984.

10. Johnson, William G., Jr.; Hill, Acquilla S.; and Eichmann, Otto: **High Reynolds Number Tests of a NASA SC(3)-0712(B) Airfoil in the Langley 0.3-Meter Transonic Cryogenic Tunnel.** NASA TM-86371, June, 1985.

11. Mineck, Raymond E.; and Lawing, Pierce L.: **High Reynolds Number Tests of the NASA SC(2)-0012 Airfoil in the Langley 0.3-Meter Transonic Cryogenic Tunnel.** NASA TM-89102, July 1987.

12. Lawing, P. L.: **The Construction of Airfoil Pressure Models by the Bonded Plate Method: Achievements, Current Research, Technology Development and Potential Applications.** NASA TM 87613, September, 1985.

13. Wigley, D. A.: **Technology For Pressure Instrumented Thin Airfoil Models.** NASA CR 3891, Contract NAS1-17571, May 1985. (Phase I SBIR)

14. Wolf, S.W.D.; and Ray, E.J.: **Highlights of Experience with Flexible Walled Test Section in the NASA Langley 0.3-meter Transonic Cryogenic Tunnel.** AIAA Paper 88-2036, May 1988.

15. Rebstock, R.; and Lee, E.E., Jr.: **Capabilities of Wind Tunnels with Two Adaptive Walls to Minimize Boundary Interference in 3-D Model Testing.** Paper given at the *Transonic Symposium*, NASA Langley, April 19-21, 1988, 26 pp. NASA CP 3020-Vol. 1, Part 2.

16. Honaker, William C., and Lawing, Pierce L.: **Measurements In The Flow Field Of A Cylinder With A Laser Transit Anemometer And A Drag Rake In The Langley 0.3-m Transonic Cryogenic Tunnel.** NASA TM-86399, April 1985.

17. Boyden, R. P.; Johnson, W. G., Jr.; and Ferris, A. T: **Aerodynamic Force Measurements With a Strain-Gage Balance in a Cryogenic Wind Tunnel.** NASA TP-2251, December 1983.

18. Cole, Stanley R.: **Exploratory Flutter Test In A Cryogenic Wind Tunnel.** NASA TM-86380, February 1985.

19. Boyden, R. P., and Johnson, W. G., Jr.: **Results of Buffet Tests in a Cryogenic Wind Tunnel.** NASA TM-84520, Sept. 1982.

20. Gartrell, L. R.; Gooderum, P. B.; Hunter, W. H. Jr.; and Meyers, J. F. **Laser Velocimetry Technique Applied to the Langley 0.3-Meter Transonic Cryogenic Tunnel.** NASA TM-81913, April, 1981.

21. Snow, W. L.; Burner, A. W.; and Goad, W. K.: **"Seeing" Through Flows in Langley's 0.3-Meter Transonic Cryogenic Tunnel.** NASA CP-2243, pp. 133-147, March 25-26, 1982.

22. Rhodes, D. B., and Jones, S. B.: **Flow Visualization in the Langley 0.3-Meter Transonic Cryogenic Tunnel and Preliminary Plans for the National Transonic Facility.** NASA CP-2243, pp. 117-132, March 25-26, 1982.

23. Burner, A. W., and Goad, W. K.: **Flow Visualization in a Cryogenic Wind Tunnel Using Holography.** NASA TM-84556, November 1982.

24. Snow, W. L., and Burner, A. W., and Goad, W. K.: **Image Degradation in Langley 0.3-m Transonic Cryogenic Tunnel.** NASA TM-84550, November, 1982.

25. Snow, W. L.; Burner, A. W.; and Goad, W. K.: **Improvement in the Quality of Flow Visualization at the Langley 0.3-m Transonic Cryogenic Tunnel.** NASA TM-87730.

26. Johnson, C. B.; and Stainback, P. C.: **A Study of Dynamic Measurements Made in the Settling Chamber of the Langley 0.3-Meter Transonic Cryogenic Tunnel.** Paper 84-0596, Presented at the AIAA 13th Aerodynamics Testing Conference, San Diego, California, March 5-7, 1984.

27. Stainback, P. C.; Johnson, C. B.; and Basnett, C. B.: **Preliminary Measurements of Velocity, Density and Total Temperature Fluctuations in Compressible Subsonic Flow.** Paper 83-0384 presented at the AIAA 21st Aerospace Sciences Meeting, Reno, Nevada, January 10-14, 1983.

28. Vakili, A. D., and Wu, J. M.: **Direct Measurement of Skin Friction With A New Instrument.** International Symposium on Fluid Control and Measurement, Volume 2, pp 875-880. Tokyo, Japan, 1985.

29. Harvey, W. D., and Bobbitt, P. J.: **Some Anomalies Between Wind Tunnel and Flight Transition Results.** AIAA Paper No. 81-1225 presented at the AIAA 14th Fluid and PlasmaDynamics Conference, June 23-25, 1981, Palo Alto, CA.

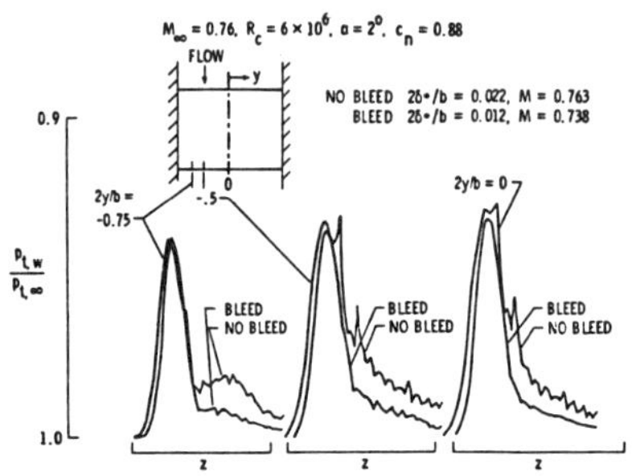

Figure 1.- Pitot surveys at three spanwise stations in an airfoil wake with and without sidewall boundary layer bleed, M = 0.76, R_c=6x10^6, alpha = 2^0, c_n=0.88.

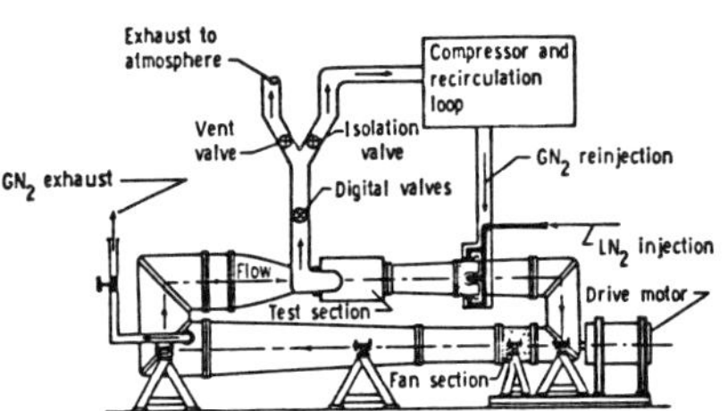

Figure 2.- Cryogenic compressor linked to the 0.3-m Transonic Cryogenic Tunnel.

Figure 3.- Pressure channels cut into opposing halves prior to brazing together to form a bonded plate model.

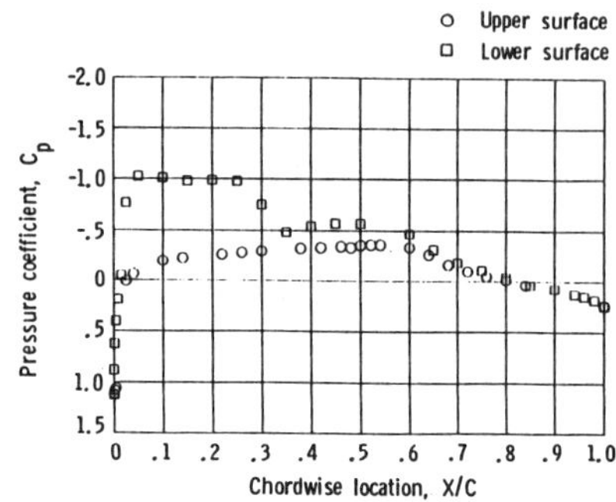

Figure 4a.- Pressure distributions on a 12% symmetrical supercritical airfoil; positive angle of attack, lift coefficient of 0.31.

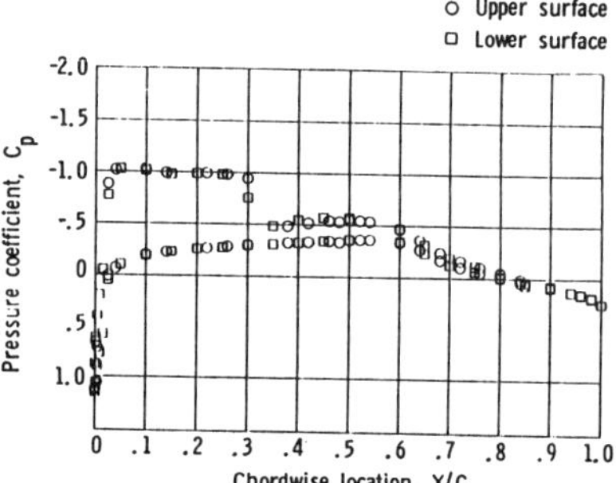

Figure 4b.- 12% symmetrical supercritical airfoil; negative angle of attack, lift coefficient of -0.31 superimposed on data of figure 4a.

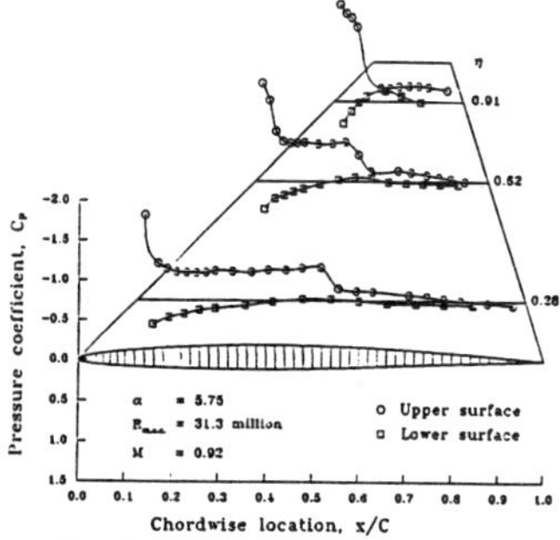

Figure 5.- Pressure distributions on the X29 canard.

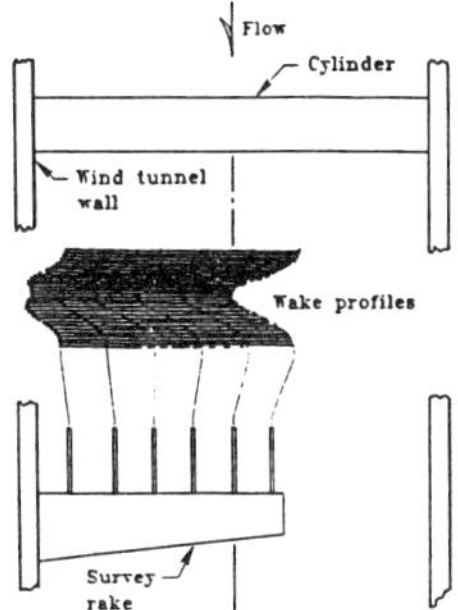

Figure 6.- Three dimensional wake topology for a transverse circular cylinder.

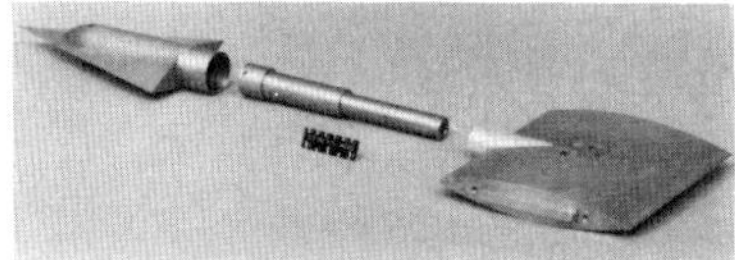

Figure 7.- Three dimensional model and support used in balance evaluation tests.

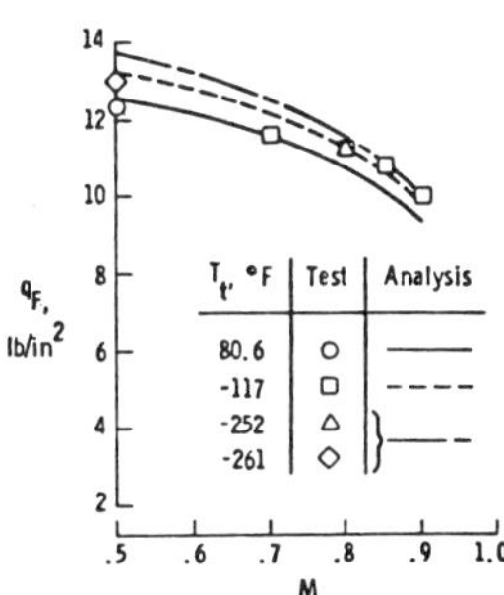

Figure 8.- Experimental and calculated flutter results.

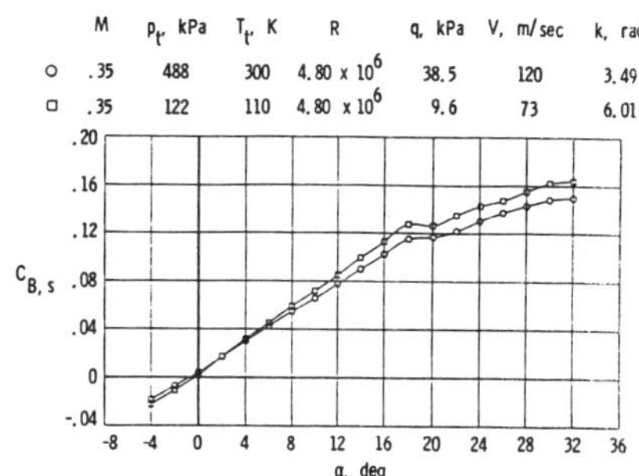

Figure 9a.- Steady and dynamic load characteristics for a sidewall mounted delta wing. Steady wing-root bending moment coefficient.

Figure 9b.- Dynamic wing-root bending moment coefficient for two values of the reduced frequency parameter.

Figure 10.- Oscillating airfoil model shown mounted in 0.3-m TCT test section.

76   Pierce L. Lawing

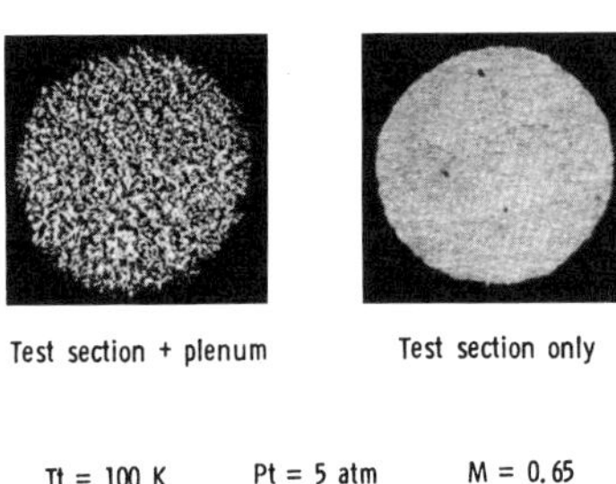

Figure 11.- Optical path through the plenum and
test section before and after isolation of the beam
from plenum disturbances.

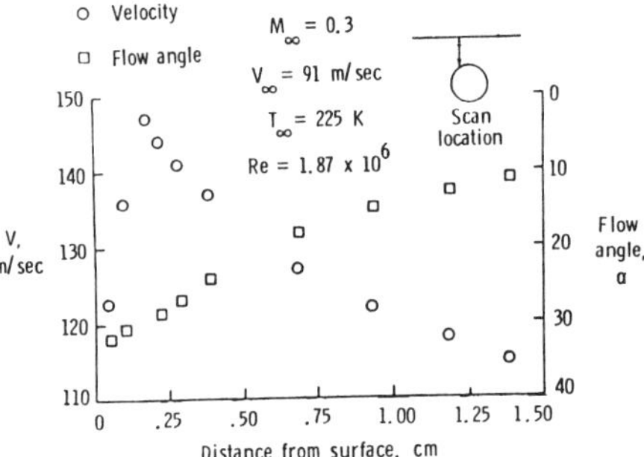

Figure 12.- Velocity and flow angle as a function of
vertical distance from the surface along the indicated
scan line.

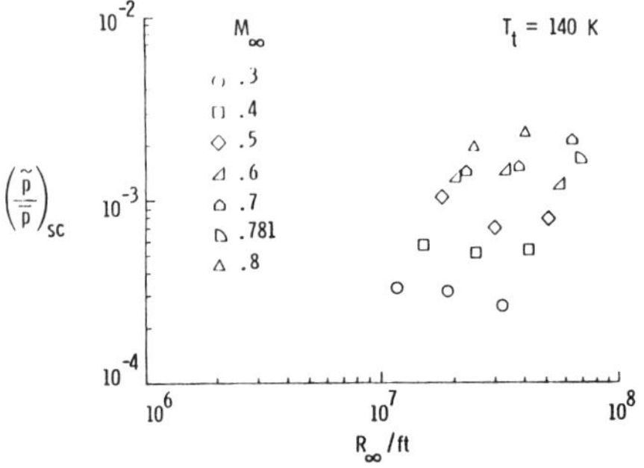

Figure 13a- Fluctuating pressure data from the
settling chamber at cryogenic conditions of 140 K.
Pressure fluctuations in the settling chamber.

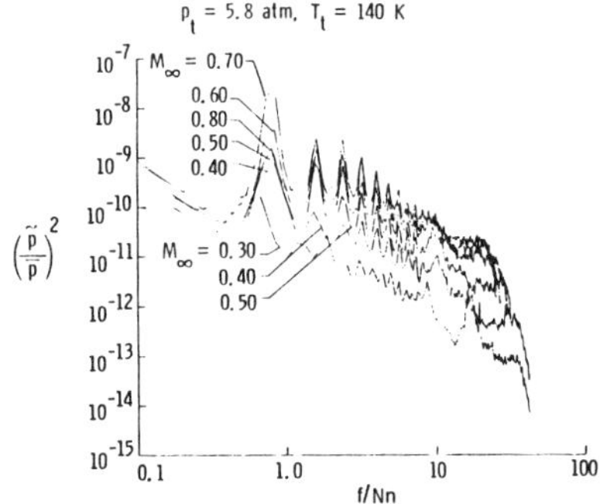

Figure 13b.- The spectra of pressure fluctuations in
the settling chamber for a dimensionless frequency
and a pressure of 5.8 atm.

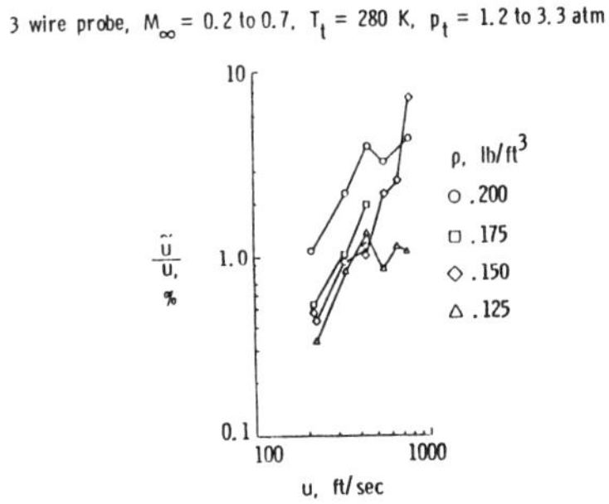

3 wire probe, $M_\infty = 0.2$ to 0.7, $T_t = 280$ K, $p_t = 1.2$ to 3.3 atm

Figure 14a.- Fluctuating quantities measured in the
test section at a total temperature of 280 K.
Normalized velocity fluctuations as a function of
free stream velocity for a range of densities.

3 wire probe, $M_\infty = 0.2$ to 0.7, $T_t = 280$ K, $p_t = 1.2$ to 3.3 atm

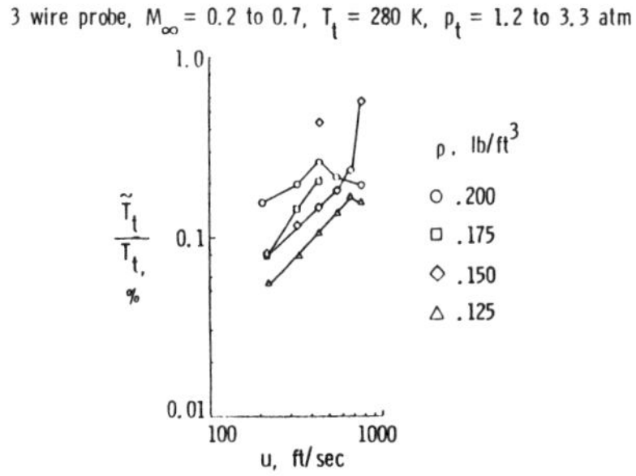

Figure 14b.- Normalized total temperature
fluctuations as a function of free stream velocity for
a range of densities.

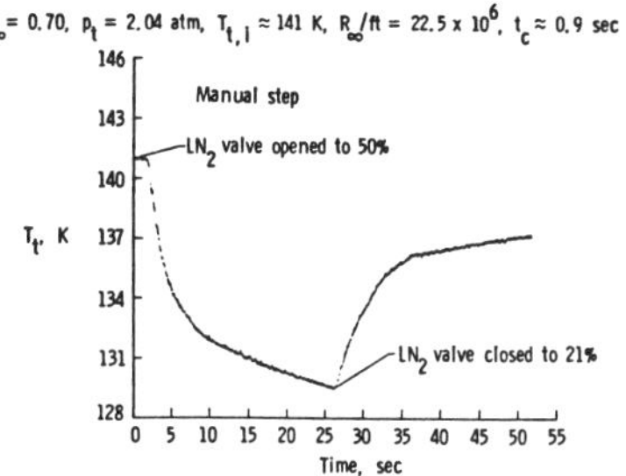

Figure 15.- Settling chamber total temperature as a function of time after a sudden input of liquid nitrogen.

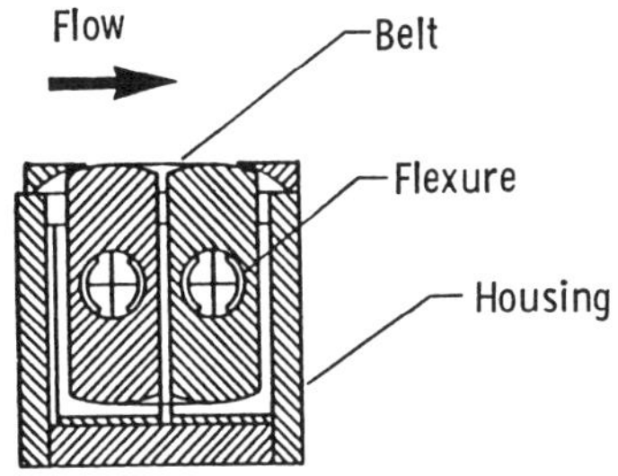

Figure 16.- Cross section of skin friction balance.

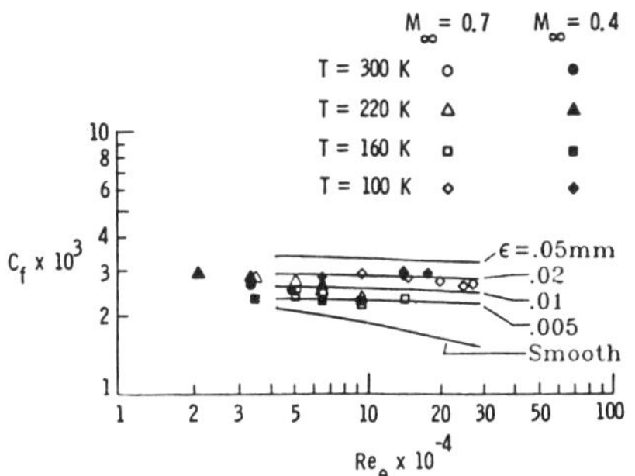

Figure 17.- Measured skin friction coefficient as a function of wall Reynolds number based on momentum thickness.

Figure 18.- Liquid nitrogen cooled airfoil model.

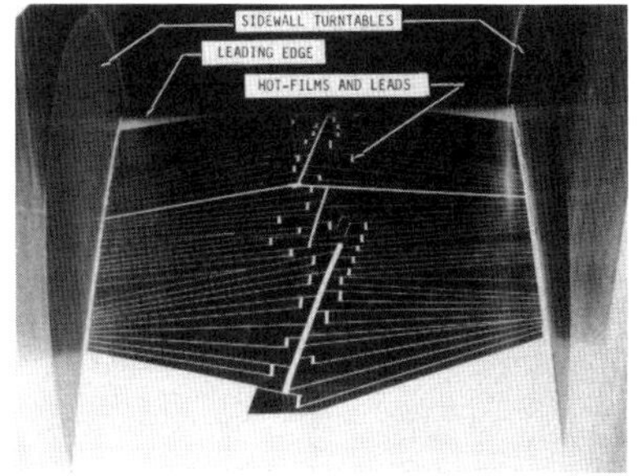

Figure 19.- Airfoil model instrumented with hot film gages shown mounted in 0.3-m TCT test section.

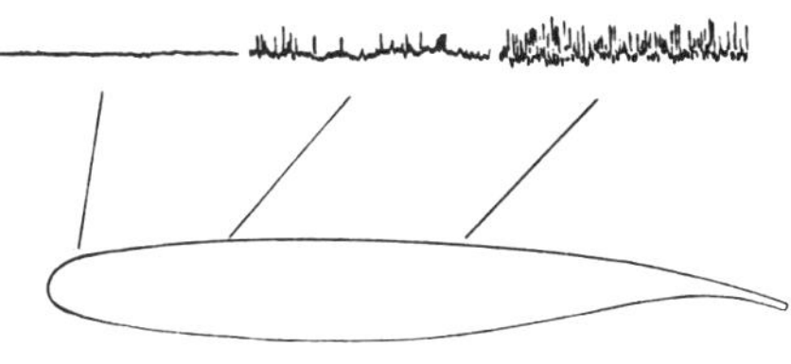

Figure 20.- 14% supercritical airfoil outline and sample fluctuating pressure traces.

HIGH-REYNOLDS-NUMBER TEST REQUIREMENTS IN LOW-SPEED AERODYNAMICS

D. M. Bushnell and G. C. Greene
NASA Langley Research Center
Hampton, VA 23665–5225

ABSTRACT

Presentation summarizes research needs in those areas of low-speed aerodynamics to which an ultra-high-Reynolds-number liquid helium facility could make a major contribution. Research areas discussed include high-Reynolds-number experimental requirements for: (a) wake vortex hazard, physics and alleviation (control of vortices) (b) supermaneuverability via body vortex control (control by vortices), and (c) high-lift device development and evaluation. Although compressibility effects are present in all of these low-speed aerodynamic application areas, observed differences between high-Reynolds-number flight and low-Reynolds-number wind-tunnel results are currently ascribed primarily to Reynolds number effects, and considerable insight into these discrepancies could be obtained in an (incompressible) liquid helium facility.

INTRODUCTION

The current shortfall in low-speed, high-Reynolds-number testing capability has a major impact on several major national programs and the expeditious solution of several critical national problems. Available low-speed Reynolds numbers (based on model length) are on the order of 30×10^6 for water tunnels and 80 to 100×10^6 for tow tanks. The cryogenic (nitrogen) National Transonic Facility (NTF) at NASA Langley is unique in providing full-scale, low-speed Reynolds-number capability $(0(10^9))$. This facility, constructed at a cost of $0(\$100 \times 10^6)$ and with a replacement cost factors of 2 to 3 greater, is not currently a high productivity device, is heavily scheduled, and is still beset with the usual "teething" problems associated with a relatively new and unique high-technology apparatus. Therefore testing availability in the NTF, even for critical national programs such as advanced technology submarine development, consists of relatively short test windows at intervals of several years. What is urgently required is a low-speed, high-Reynolds-number facility that is inexpensive enough so that perhaps several could be eventually produced and with improved productivity/capability. A liquid helium (HeI) facility would uniquely fulfill this requirement. An additional feature of such a facility is the ready use of superconducting magnetic suspension to obviate support interference and allow incorporation of propulsive devices and measurement of dynamic body motion and control effects.

The ready availability of low-speed, high-Reynolds-number testing capability can have several, potentially revolutionary, impacts on national programs such as submarine and surface ship development and improved airport utilization via wake vortex hazard reduction. This impact would be two-fold, the ability to reduce risk, time, and treasure associated with evolutionary improvements and, perhaps most importantly, the capability of evaluating, at full-scale conditions, the performance of revolutionary concepts. Currently, the almost unmanageable risks associated with extrapolation from low-Reynolds-number experiments, along with the inaccuracies of computational fluid dynamics (CFD)/turbulence modeling, severely inhibits the evaluation and adoption of revolutionary concepts for submarines, aircraft, and surface ships. Such a facility could also be designed to address, in addition to body and propulsive performance, the critical submarine acoustic performance and minimization issues.

The purpose of the present contribution is to briefly summarize some high-Reynolds-number research requirements in low-speed aerodynamics which could be addressed, to a significant extent, in a liquid helium tunnel. Although compressibility effects are present in all of these areas, much of the existing discrepancies between flight and tunnel results are currently ascribed to Reynolds number effects and, therefore, addressable via an "incompressible" high-Reynolds-number (liquid helium I) facility. Particular low-speed aerodynamic research areas addressed include: (a) wave vortex hazard reduction, (b) vortex control and diagnostics for maneuvering fighter aircraft, and (c) performance of high-lift devices. In addition to these application areas, there are additional, more generic, high-Reynolds-number aerodynamic research requirements such as the fundamental structure/characteristics of turbulence in attached, separated, and (organized) vortical shear flows.

WAKE VORTEX HAZARD REDUCTION

The need for some means of addressing the wake vortex hazard was first identified as a serious issue in connection with the emergence of the heavy "wide-body" aircraft in the late '60's — early '70's (747, DC-10, L1011). During landing/takeoff the wing tip vortices from such aircraft are sufficiently strong and long-lived to endanger smaller aircraft flying through them. Typically the "danger zone" associated with the tip vortices from wide-body aircraft extends for the order of 7 miles or more downstream. Hazards to lighter aircraft encountering such vortices include extreme structural loading, imposed rolling moments, and/or sudden loss of altitude, depending upon the relative orientation of the flight paths. These wing vortices are, during the high-lift condition typical of landing and takeoff, actually initially an extensive collection of individual vortex elements which emanate from the edges of the various flaps, etc., as well as from the wing tips themselves. These vortex arrays usually interact downstream to form strong (oppositely rotating) vortical entities from each wing which, in turn, eventually interact downstream and ultimately breakup/dissipate, etc. The hazard associated with these wake vortices occurs during the extensive pre-breakup/dissipation region (ref. 1).

There are two basic approaches to treating the wake vortex hazard problem: (1) detect and avoid (historically favored by the FAA) and (2) ameliorate/control (historically favored by NASA). The initial period of research as this problem (early to mid-'70's) was unable to yield satisfactory systems for either of these approaches and a default position was adopted — lengthen the distance between aircraft in the terminal area to allow for "natural dissipation" of the vortices. Such natural dissipation is highly variable depending upon specific aircraft configuration and atmospheric conditions/turbulence and, therefore, a conservative approach had to be adopted, resulting in a significant loss in airport and aircraft productivity. The tremendous growth in the U.S. air transportation system, coupled with the costs, etc., associated with airport construction has refocused attention on the "solution" of the wake vortex hazard problem as a way of improving current airport usage and thereby delaying the construction of additional airports. The air traffic control system upgrades currently in the pipeline would allow closer aircraft spacing if the wake vortex hazard problem could be successfully addressed. Probably the optimal method of approach to this problem would be through control of vortex genesis and near field behavior. It should be noted that such control does not imply a reduction in the vorticity produced by the aircraft, a certain amount of which is necessary to produce lift. What is espoused is the reorganization/disorganization of this vorticity into "kinder, gentler" flow gradients.

The NASA research of the 1970's resulted in several approaches and inventions which could, and did, disable the high flow gradient portions of the wake vortex. This research was carried out in both flight and a collection of ground facilities including wind tunnels and tow tanks. These control approaches, originally developed in the ground facilities, included: (a) vortex segmentation, subsequent alteration of the rollup process, (b) tip turbines/engines/winglets, (c) mass injection at the tip, (d) direct (towed) turbulence generators, and (e) oscillatory (e.g., via flap/spoiler movements) flow inputs. The results of much of this ground-based research, as well as the "unmodified aircraft" tare case obtained from ground facilities were then compared with corresponding flight data (e.g., ref. 1). The results of such ground-flight comparisons were almost universally disappointing with the ground results being greatly optimistic. The ground data simply could not be relied upon to represent the flight case, and the various alleviation devices had to be reoptimized all over again in flight, a horrendously expensive procedure which led, in due course, to the cancellation of the research program and the present state-of-affairs in terms of compromised airport productivity.

A probable cause of these discrepancies between flight and ground data was described in reference 2, a large disparity in vortex Reynolds number (on the order of two orders of magnitude). The much lower Reynolds numbers of the ground facilities has two major consequences: (1) much greater viscous losses within the vortex leading to significantly earlier dissipation and (2) relative absence of strong turbulence influences upon the vortex. Obviously shear turbulence in a vortex is altered, to first order by Rayleigh stabilization due to <u>flow curvature</u> and this is a major reason for the long-lived nature of such flows. The presence, at high Reynolds numbers, of an annular region of turbulent flow evidently alters the vortex structure to first order. As an example, vortex velocity field measurements in flight indicate a multicell nature of the longitudinal component, a behavior which is not generally observed in ground experiments.

Therefore, the solution of the wake vortex hazard problem, an important national issue with huge (multibillion) financial ramifications, via vortex control/minimization hinges upon the availability of a high-Reynolds-number ground facility suitable and available for parametric, "Edisonian" experimental studies. The size and limited availability and productivity of the NTF (the only current candidate facility) necessitates the development of other low-speed, high-Reynolds-number ground capability. The proposed liquid helium tunnel uniquely fulfills this requirement. It should be noted that a similar Reynolds number simulation requirement occurs for the submarine trailing vortex problem (ref. 3).

VORTEX DIAGNOSTICS AND CONTROL FOR "SUPERMANEUVERABILITY"

The ability to turn fast and "point and shoot" is a key ingredient to fighter success and survivability (ref. 4).   To accomplish this, the aircraft must be controllable and have reasonable performance at high angles of attack. Several of the initial generation of jet fighters, through the "century series," employed direct boundary-layer control (blowing) to keep the flow attached and thereby enhance maneuverability.   The research connected with the "Concorde" SST identified the improvement of "vortex lift" and more recent fighters utilize longitudinal vortex generation via "leading-edge extension" (LEX) or other vortex generation techniques to alter and control the separated flows endemic to high angle-of-attack aerodynamics (e.g., ref. 5). Unfortunately, these vortices themselves can become unstable/breakdown giving rise to highly nonlinear aerodynamics and enhanced structural loading. Therefore, this problem area involves both control by vortices and control of vortices.

The conventional wisdom regarding the behavior of such body vortices is that their behavior is Reynolds number insensitive. This conclusion is based upon extensive data, both ground and flight, for vortex burst location (e.g., ref. 5). However, the physics of the situation suggests that this vortex bursting behavior is dictated primarily by the body (largely inviscid) pressure field, in particular adverse pressure gradients.   However, as reference 6 points out, the detailed growth and behavior of these body vortices, other than their (pressure-gradient dictated) burst location, are in fact strong functions of Reynolds number, including axial velocity profiles and resultant body forcefield (e.g., influence(s) upon flow separation).   Therefore, as in the wake vortex hazard case, ground simulation of detailed body vortex behavior and control requires a high-Reynolds-number facility, e.g., the liquid helium tunnel.

HIGH-LIFT SYSTEM PERFORMANCE

The low-speed, high-lift capability required for aircraft landing and takeoff is a critical design issue, especially for long-haul transport aircraft. The current "approach of choice" for such systems includes disparate leading-edge strakes/slats and various forms of trailing-edge flaps. The basic issue concerning the performance of such devices is the transition behavior over the leading-edge device portion of the high-lift system and the subsequent influence upon flow separation occurrence on the wing. The ground simulation of such leading-edge transition behavior requires Reynolds numbers, at low-speeds, currently unavailable except in NTF. The problems with NTF availability and feasibility for such research have already been mentioned in connection with the wake vortex hazard discussion herein. An additional problem with the NTF for the high-lift problem is its high dynamic pressure which makes design and survivability thin leading devices problematical. The low-Reynolds-number ground facilities currently available (except for NTF) usually provide pessimistic estimates for system performance due to late transition and enhanced flow separation. What is required is a high-Reynolds-number, <u>low-dynamic pressure</u> ground facility, again the proposed liquid helium tunnel uniquely fulfills this requirement.

CONCLUDING REMARKS

Paper briefly summarizes some of the nation's low-speed aerodynamics requirements for a high-Reynolds-number liquid helium tunnel. Such a facility could uniquely address critical Reynolds number questions/requirements in the areas of wake vortex hazard reduction and improved fighter maneuverability and high-lift systems. All of these aerodynamic transport problems also entail compressibility influence to various degrees, the absence of which in a low-speed liquid helium tunnel would somewhat compromise the results. However, the current Reynolds number shortfall is such an overriding issue that the liquid helium facility would contribute, in a major way, to their solution. Typical full-scale test requirements for such aerodynamic application include vortex Reynolds numbers of 10^7 on a 3–D semi-span wing with aspect ratio 10 and lift coefficient on the order of 3, a fighter body length Reynolds number on the order of 100×10^6 at up to $90°$ angle of attack and "multielement" high-lift wing with aspect ratio 10, lift coefficient of 3, and chord Reynolds number on the order of 30×10^6.

REFERENCES

1. Wake Vortex Minimization, NASA SP409, 1977. Proceedings of a symposium held in Washington, DC, Feb. 25–26, 1976.

2. Iversen, J. D.: Correlation of Turbulent Trailing Vortex Decay Data. AIAA J. Aircraft, V. 13, May 1976, pp. 338–342.

3. Lee, H.; and Schetz, J. A.: Experimental Results for Reynolds Number Effects on Trailing Vortices. AIAA J. Aircraft, V. 22, No. 2, Feb. 1985, pp. 158–160.

4. Dorn, M.: Aircraft Agility: The Science and the Opportunities. AIAA Paper 89–2015 presented at the AIAA/AHS/ASEE Aircraft Design, Systems and Operations Conference, Seattle, WA, Jul. 31 — Aug. 2, 1989.

5. Erickson, G. E.: Vortex Flow Correlation. AFWAL-TR-80-3143, Jan. 1981.

6. Wortman, A.: On Reynolds Number Effects in Vortex Flow over Aircraft Wings. AIAA Paper 84–0137 presented at the AIAA 22nd Aerospace Sciences Meeting, Reno, NV, Jan. 9–12, 1984.

FLOW VISUALIZATION

Leonard M. Weinstein
NASA Langley Research Center
Hampton, VA · 23665-5225

ABSTRACT

Flow visualization techniques are briefly reviewed. Both
qualitative and quantitative methods are examined, with
special emphasis given to techniques applicable to liquid
helium flows. Focusing schlieren, particle image velocimetry,
and holocinematography are described in some detail. The last
of these in particular can give the time-varying three-
dimensional velocity field.

INTRODUCTION

The earliest studies of fluid dynamics consisted of
observations of nature. By studying the flight of birds and
by observing water swirl around obstacles for example, at
least some of the fluid physics was observed. Later,
controlled experiments, where flow visualization was combined
with quantitative measurements (pressure, temperature, etc.),
allowed more details of the flow to be understood. Flow
visualization remained an important tool even when results
were limited to qualitative observation, because it rendered
certain properties of a flow field directly accessible to
visual perception. When advances in both flow visualization
and quantitative measurements were combined with mathematical
models of flow, understanding grew further. Techniques such
as the hot-wire anemometer and laser Doppler velocimeter
allowed time-varying flows to be measured at one or several
points. Application of Taylor's Hypothesis allowed the time-
varying measurements at a point to be related to the spatial
details of flow structure. For highly complicated flows,
better understanding could be obtained by combining time-
varying point measurements with flow visualization to generate
a correlated picture of the flow field. References 1 and 2
give good reviews of many of the most useful techniques for
complex flows. For higher speed flows, only two-dimensional
slices (smoke or vapor screen with laser light sheet) or
integrated two-dimensional pictures (schlieren,
interferometry, etc.), could be obtained at one time. This
has limited our ability to understand many of these flows.

Generally, the first approach to understanding a complex
flow field is to use flow visualization to convey the broad
picture to the researcher. This flow visualization might

include off-body as well as model surface flow
visualization. This qualitative approach often provides
fundamental insight into the physics of the flow; however, it
does not usually provide adequate details for the development
or validation of analysis techniques. The next step usually
consists of detailed measurements made at selected point
locations in the flow field or on the model. For complicated
flows, especially unsteady ones like wakes or turbulent
boundary-layer flows, more global observations and
measurements are needed.

Experimental measurements of sufficient spatial
and temporal resolution to be useful for understanding
complex flows have been described as Quantitative Flow
Visualization (QFV). QFV are flow-field measurement
techniques where sophisticated sensors and digital image
processing are used to extract the required data from the flow
field of interest. Recent advances in hardware and software
have made image processing of flow visualization photographs
possible in near real time. Two- and three-dimensional data
can be obtained and processed to examine flow details.

The present article examines flow visualization
techniques with emphasis on those that might be usable to
study liquid helium flow. The use of liquid helium as a test
fluid allows large Reynolds numbers to be obtained on small
models and at low velocities.

FLOW VISUALIZATION TECHNIQUES

The majority of flow visualization techniques can be put
into two groups (Ref. 3). Tracer techniques require the
addition of a foreign material such as smoke in air or dye in
water and optical techniques require variations of the
refractive index of the flowing medium. Quantitative
evaluation of most optical measurements is difficult since the
record of the refractive index distribution is an integrated
value along the path of the light. Table I gives a partial
list of examples of flow visualization techniques.

TABLE I Flow measurement techniques

Tracer Techniques:	Water	-	Dye
			Bubbles
			Chemicals
			Particles
	Air	-	Smoke
			Helium/Soap Bubbles
			Tufts
			Particles
			Surface Oil Flow
Optical Techniques:			Schlieren
			Shadowgraph
			Interferometry

The use of lasers greatly improves the qualitative and quantitative visualization techniques possible. The ability to obtain high brightness points and thin sheets along with short exposure and even holographic images allows a wide variety of techniques to be used (Ref. 4). If discrete particles are tracked either by use of LDV, high-speed photography or by using electronic cameras, the velocity can be obtained at points over areas, and with holography (or rapidly scanned light sheets) velocity information can be obtained over volumes.

Table II gives a partial list of techniques which give quantitative flow-field measurements for three-dimensional flows. The first group is for steady or slowly-changing flow fields. The second group attempts to obtain full three-space and time information, but tends to be limited to fairly low-speed flows by the information collection rate.

TABLE II Quantitative flow-field measurements

Methods to Measure 3-Space for Steady Flows:
    Stereoscopic Photography of Tracers
    Scanning LV
    Scanning Light Sheet
    Focusing Schlieren
    Fluroluminescent Velocimeter

Methods to Measure 3-Space and Time:
    Sterescopic Movies of Tracers
    Double Pulse Holography Movies
    Rapid Scanning Light Sheet
    Holocinematographic Velocimeter

The techniques which would be usable for liquid helium flows are discussed in the next sections.

MEASUREMENTS IN LIQUID HELIUM FLOW

The purpose of making measurements in liquid helium flows is to obtain very high Reynolds number flows at low physical speeds in small scale facilities. Both qualitative and quantitative measurements would help understand the flow physics of highly complex time-varying, three-dimensional flow fields.

In helium flows, temperature and thus density variations could exist in the flow and optical methods could be used to examine flow structures. One variation of schlieren, called focusing schlieren (Ref. 5), could be used to examine two-dimensional slices of the flow. Figure 1 shows the optical system required (from unpublished work by Leonard M. Weinstein).

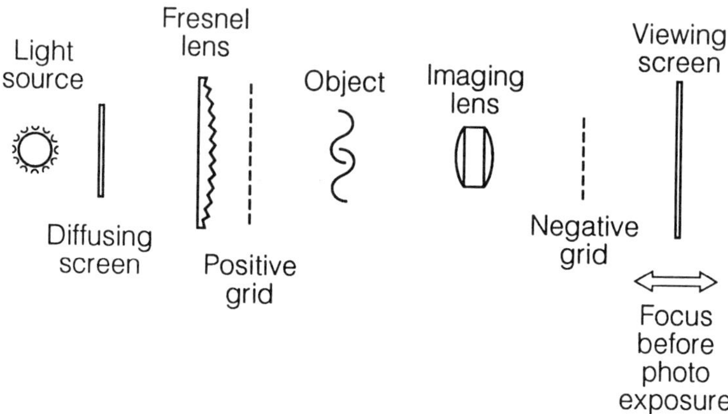

Figure 1.   Focusing schlieren system with field lens.
(Landreth, et al.)

By rapidly moving the image plane location, different depth
slices are examined. If it were possible to scan rapidly
enough, the full three-dimensional flow might be examined and
it should be possible to use the difference between adjacent
planes to get local quantitative results (this has not yet
been demonstrated). If the setup of Figure 2 were used (from
unpublished work by Leonard M. Weinstein), full three-
dimensional schlieren photos could be obtained in single-
exposure holograms.

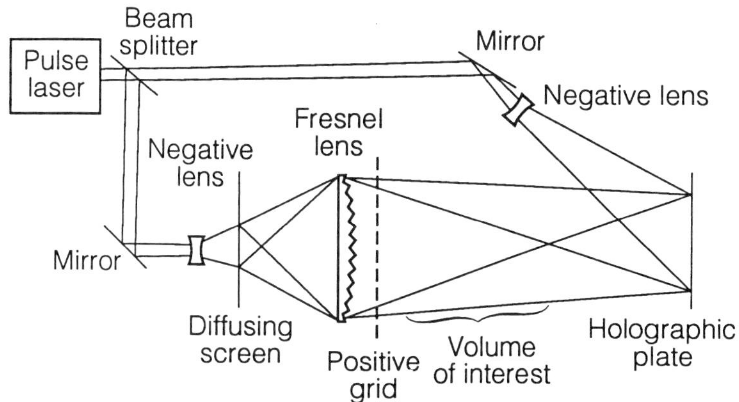

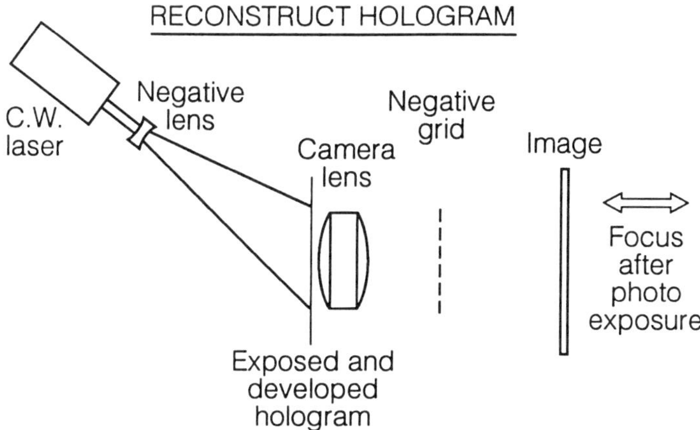

Figure 2. Three-dimensional schlieren system.

Use of high rep-rate lasers and fast film transports could allow full-time varying, three-dimensional flow fields to be examined.

The use of tracer particles in the flow would allow a variety of particle tracking techniques to be used. Two methods of visualizing the flow and obtaining velocity (Weinstein, et al., 1985) information using tracer particles will be discussed in the next sections.

PARTICLE IMAGE VELOCIMETRY

Particle Image Velocimetry (PIV) uses the movement of discrete particles to obtain the velocity field. The most commonly used technique is to illuminate a plane slice of the flow with a short intense double (or multiple) pulse of light, to record instantaneously the two-dimensional velocity vector field in the planer slice of the flow. A large number of points may be used to obtain high spatial resolution. If the plane is rapidly swept, and high-speed movies are made, the full volume may be examined. Note that even for the swept sheet, the results are still only two-dimensional. Figure 3 from Reference 6, by C. Landreth, et al., shows one setup.

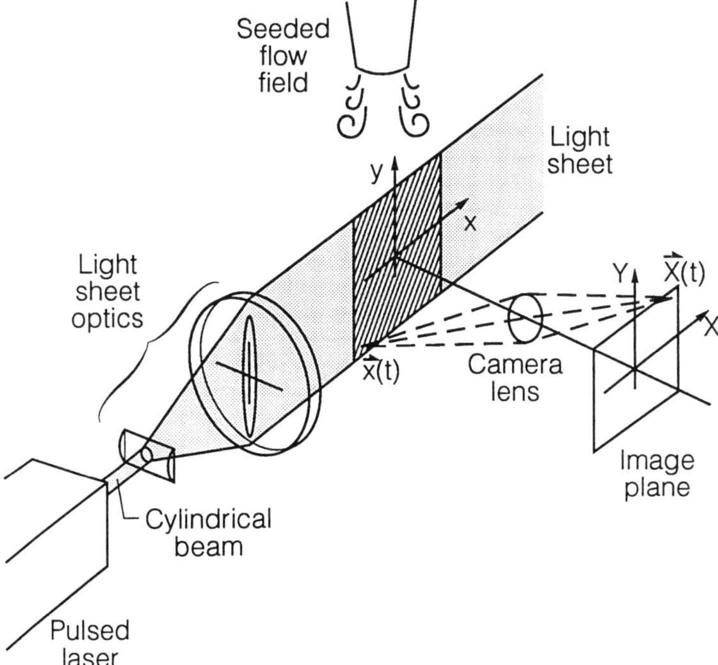

Figure 3. Light sheet and photographic system for particle
image velocimetry.(Landreth, et al.)

The large number of points requires an automatic data reduction technique be used. Adrian described two types of schemes to analyze the photographs. One, called an image compression system, achieves a substantial savings in the number of pixels that need to be handled by imaging the round interrogation spot onto a row of 1024 photodiodes by using a cylindrical lens. Images within the spot are effectively projected onto the 1 x 1024 diode array and the displacement is easily determined within the computer when no more than 2 images are present (Fig. 4).

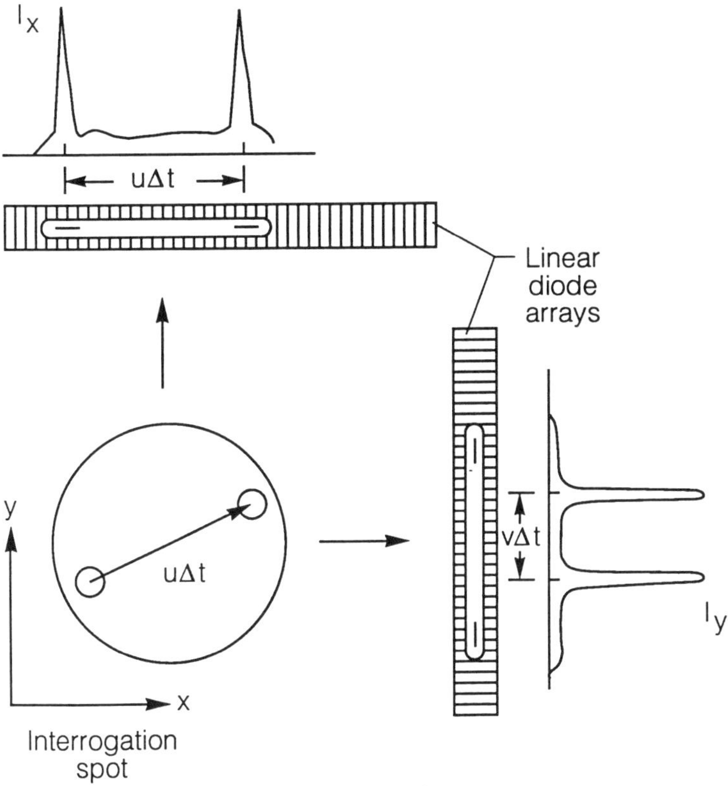

Figure 4. Orthogonal compression technique.

Displacements in the perpendicular direction are obtained
from an orthogonally oriented cylindrical lens and a
companion diode array. In this way, total data transfer
per square millimeter is 2043 pixels, although the
effective resolution is the same as1024 x 1024 pixels.
He found that this system works extremely well when the
average number of particles per interrogation spot is
small. For this case the probability of particle pairs
overlapping is negligible and each pair of images can be
assumed, with high probability, to be from the same
particle. It is possible to analyze interrogation spots
at a rate of approximately 1.2 sec per spot using a small
DEC 11/23 computer enhanced by a Skymak processor. Accuracy
has been found to be better than 0.6 percent of full scale
and it is possible to obtain measurements of several thousand
vectors within a typical field view.

More recently, Adrian upgraded his computer system
to a Microvax II enhanced by a Numerix 432 array processor.
The latter is a very powerful 30 mega-flop machine selected
for its ability to compute 256 x 256 convolutions in less
than one second. This machine can interrogate particle
image photographs containing many particle images per
interrogation spot. This is desirable because every
interrogation will then yield a velocity vector measurement
during the interrogation process, but results in some
confusion in determining the pairing of the images. The
proper pairing can be sorted out by performing a two-
dimensional spatial convolution of the interrogation
spot image field and seeking the maximum of the correlation,
as shown in Figure 5. The accuracy in an example case was
estimated to be better than 1 percent. Each frame was
analyzed from a photograph in less than 1/2-hour using the
spatial correlation technique.

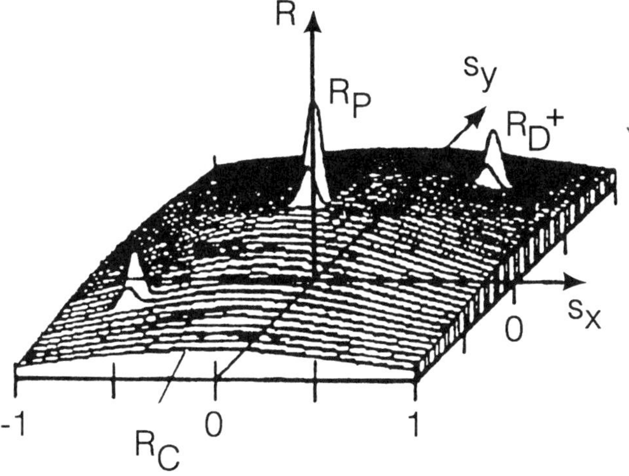

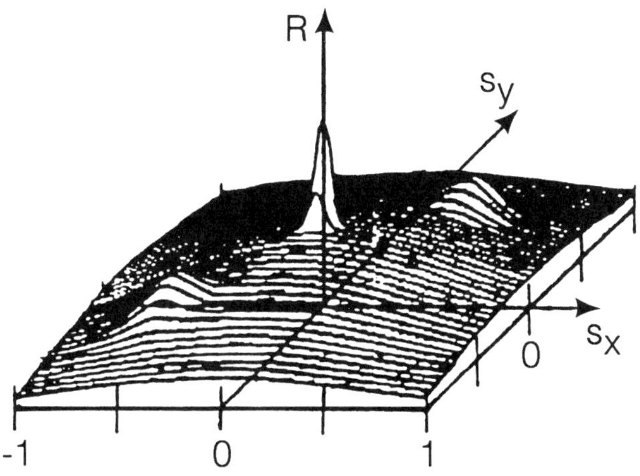

Figure 5. Convolution of the interrogation spot image. The secondary maxima locate the mean displacement of the images in the interrogation spot.

HOLOCINEMATOGRAPHIC VELOCIMETER

Another version of PIV called the Holocinematographic Velocimeter (HCV) was described in References 7 and 8. This technique used holographic movies of tracer particles to examine the flow in a low-speed water tunnel. Figure 6 provides a simple demonstration of the technique.

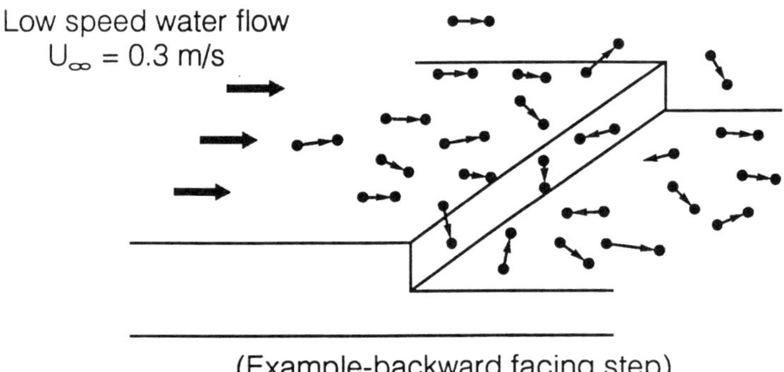

- Make holographic movie of 3-d flow with tracer particles
- First in each sequential pair of frames ➡ X, Y, Z, at t
- Second in pair is located ➡ ΔX, ΔY, ΔZ at Δt
- Data reduction for entire movie yields (in real time) V(X, Y, Z, t)

Figure 6. Determining time-varying velocity field with HCV.

High-speed (single exposure) holographic movies examine
large numbers of tracer particles in the volume being
examined. Since particle locations can be measured
accurately in directions normal to the focal axis, but
less accurately along the focal axis, (Ref. 7) a
simultaneous second view of the flow field is used to
obtain equal position accuracy in all three spatial
dimensions. Particle locations are found with an
automatic image processing system. Particle motion
determined from successive frames in the movie is
used to obtain the time-varying velocity field.
Table III gives the capabilities of the system being
developed.

TABLE III Capabilities of system being considered.

Flow velocity of 0.3 m/s in low-speed water tunnel
Tracer particles 40μ diameter
Tracers 1 mm apart; test volume 25 X 25 X 35 mm --
  2×10^4 tracers
Dual high-speed cameras at 1000 frames/sec for 1 sec --
  4×10^7 sets of coordinates
Particles move 0.3 mm/frame; position accuracy 5μ
Typical run will take about 1-1/2 weeks to reduce data
Resolution of turbulence motions down to 15^+ (wall units)

A new type of water tunnel was developed to use with
the HCV. The tunnel has a very low disturbance flow and
is less expensive to construct than continuous flow or
even intermittent flow water tunnels. The flow velocity
range and run time are limited, but sufficient for the HCV.
A sketch of the facility is shown in Figure 7.

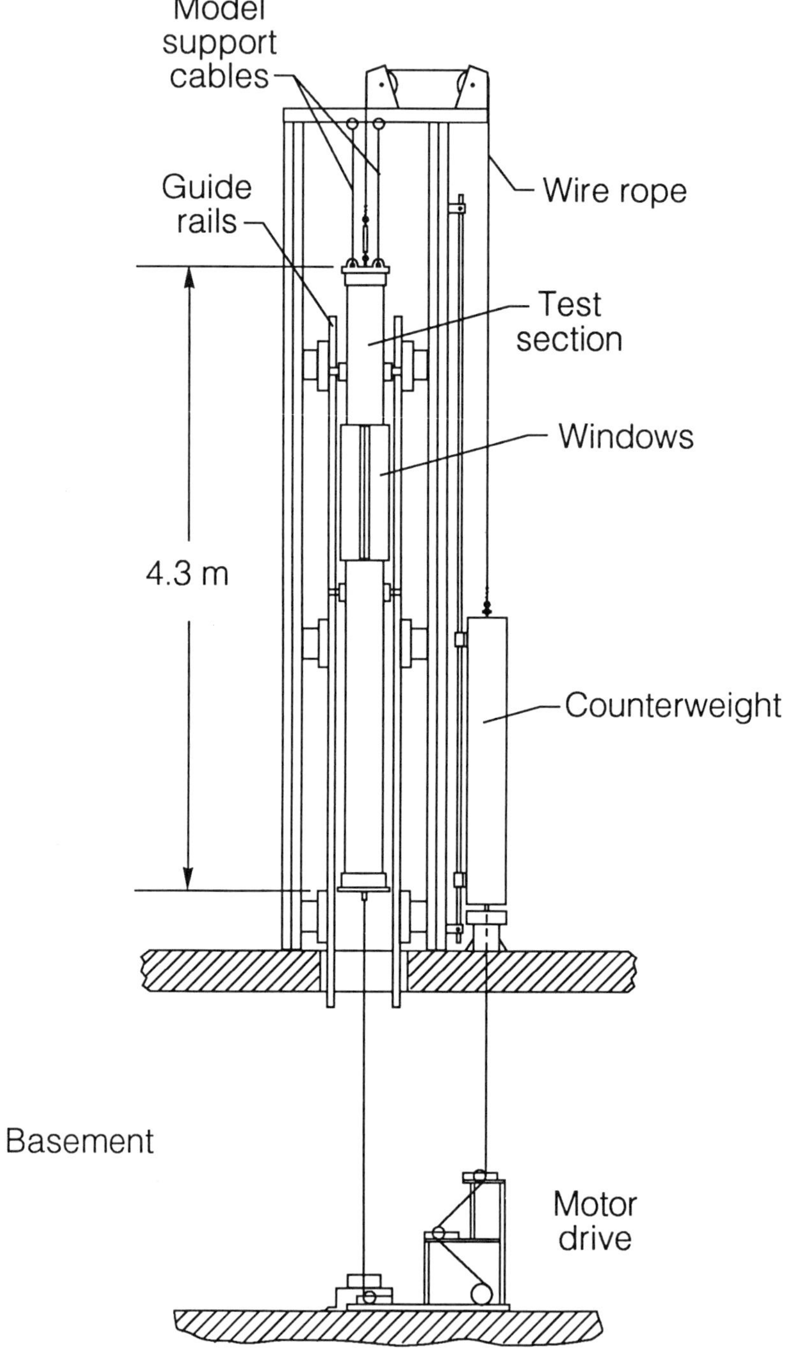

Figure 7. Vertical translated water tunnel.

The vertical test section is a 0.25 m diameter pipe 4.3 m long
which is sealed at the bottom but partially open at the top
and filled with water. Four optical quality windows 0.15 m
wide by 1 m long are mounted starting 0.7 m from the top of
the test section. An interior plastic rail guides the model
the length of the test section. The model is attached to two
support cables which are secured to the test section support
frame. The test section has linear bearings to guide it along
vertical rails. A counterweight is used along with support
cables and a drive system to obtain 2.4 m of travel at speeds
up to 1 m/s. The test section is raised to the highest
position and the model location is adjusted with the support
cables. After initial flow currents have died out, tracer
particles, which consist of 40μ Polysterene microspheres
(standard deviation 3μ), are released into the test volume.
When they are well dispersed, the test section is smoothly
accelerated to the test velocity then held at constant
velocity to allow the pipe full of water to be lowered past
the stationary model. When the tunnel windows traverse the
field of view, the holographic movies are taken.

A cross section of the tunnel is shown in Figure 8 along
with the optical setup used.

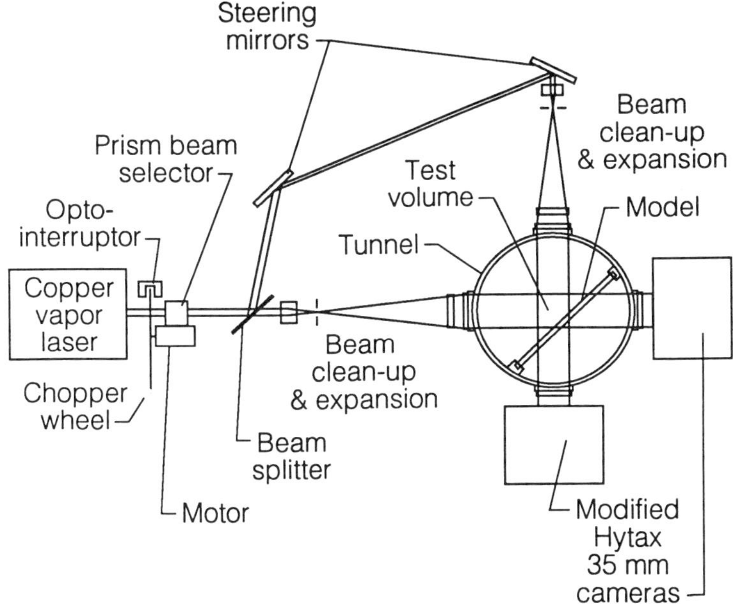

Figure 8. Optical setup and beam control for
in-line holography.
(Weinstein, et al., 1985)

A flat plate model with a glass insert is held with wire
supports and guided by the side rails inside the test
section. Use of reference marks on the glass insert allows
accurate coordinate measurements with respect to the model.
Expanded laser beams intersect in a common volume as shown and
are recorded on two high-speed film transports. A pulsed
copper vapor laser is used with the capability of obtaining
10,000 pps at about 0.3 mJ per pulse at 510.6 nm. Due to beam
splitting and masking, only about $100\mu J$ per picture are
obtained. An optical chopper selects the desired pulses.
Since 1,000 photos per second are needed, only every tenth
laser pulse is allowed to pass the optical chopper wheel. The
film used is Kodak LPF-4 which requires 4 ergs/cm^2, and which
has 800 ℓ/mm resolution. This film is high contrast and
suitable for low resolution holography. The film transports
are modified Hytex 35 mm cameras and can be used for full-
frame rates up to 2,000/sec. The cameras were modified to
give a flat frame fully 35 mm long, and the film transport
speed is set to avoid overlap of exposures. A 150 ft. roll of
film lasts only a little over 1 second at 1,000 frames per
second.

     After the holographic film is processed, it is examined
and reduced with the setup of Figure 9.

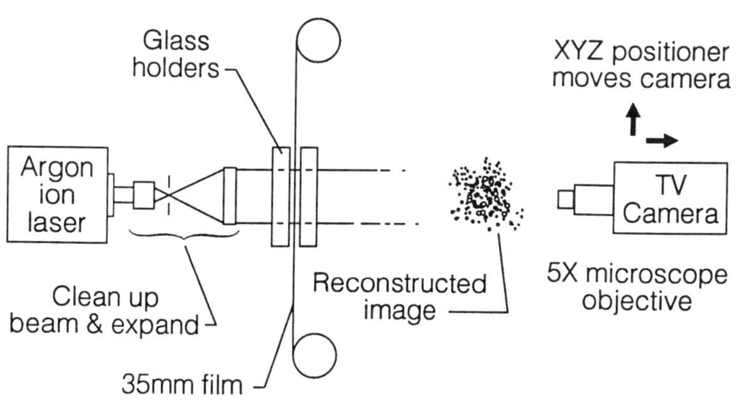

Figure 9. Reconstruction optical system.

An argon ion laser (λ = 514nm) is expanded and collimated to
cover a full frame. A real image of the tracer particles is
generated on both sides of the film, and the one on the far
side is used for data reduction. Due to the difference in
wavelength of the copper vapor laser and argon ion laser, and

also due to the difference of index of refraction of the water and air, the image has a different scale along the optical axis. This is corrected in software. The vertical and lateral scales are not changed by those effects, but may be slightly in error if either the copper vapor or argon ion lasers are not perfectly collimated. Two reference marks on the model allow scale correction.

A video camera mounted on a three-axis positioner can examine any part of the image field with a microscope objective used to obtain a magnified image. The film is positioned and glass cover plates hold the film flat and steady. The entire reconstruction setup is mounted on a vibration isolation table to obtain the required accuracy.

The image analysis system examines each hologram as a series of two-dimensional slices. Each thin parallelepiped in Figure 10 represents a video camera viewfield with 1 mm depth.

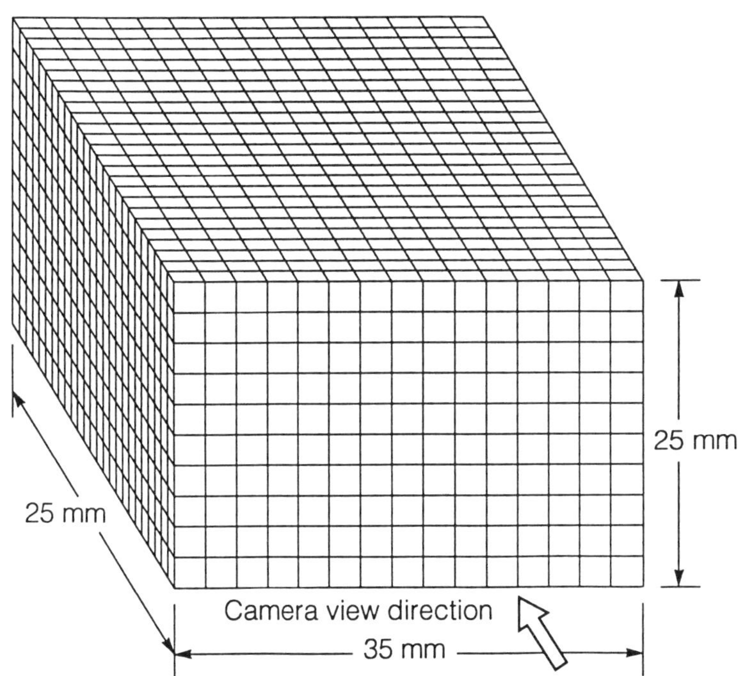

Figure 10. Test volume being examined.
(Weinstein, et al., 1985)

The volume is divided into 2.5 mm square areas. This results
in a video resolution of 5μ, which sets the system accuracy.
The camera is moved in depth and the particle stays in
reasonable focus for about 3-depth steps. New lateral and
vertical locations allow the entire volume to be examined.
Each video frame takes an average of 0.12 second to grab and
reduce to partical coordinates, so each hologram takes about 8
minutes to reduce. Even at this rate, 2000 holograms (1000
each for two views) takes nearly two weeks to reduce.

The coordinates generated are transferred to a Gould
computer to further reduce the data. The multiple views of
coordinates corresponding to different focal locations of one
tracer are reduced to one value by combining data from the two
views. Next the velocity field is determined with these
coordinates by particle tracking.

After the run is reduced to the time-varying, three-space
velocity field, the data can be utilized in the same manner as
results from a numerical turbulent simulation. The data can
be used to examine coherent/turbulent structures in non-
simple/non-attached flows. It can be used for turbulence
simulation input, turbulence modeling and flow
visualization. Many of the programs and data handling
procedures developed to examine results of computer
simulations can also be applied to the experimental data base
obtained with the HCV. This would allow early utilization of
the data obtained with the HCV.

The HCV appears to be uniquely capable of obtaining data
with a time and 3-space resolution that approaches a full
real-time measurement of complex flows. Although the current
version is limited to low-speed (e.g. water and liquid helium)
flows, this is sufficient to examine a large number of complex
flow problems.

CONCLUSIONS

Flow visualization techniques were briefly reviewed with
special emphasis given to techniques applicable to liquid
helium flows. Three techniques were selected that are capable
of obtaining qualitative and quantitative measurements of
complex three-dimensional flow fields.

Focusing schlieren can be used to examine the density
field if sufficient temperature variations are present. It
may be possible to obtain quantitative variations by using
techniques similar to those used in Tomography.

Two-dimensional PIV can give high resolution slices of
the velocity field, and if scanned rapidly, can examine a
volume. The result, however, is still two-dimensional for
each sheet.

The HCV appears to be uniquely capable of obtaining full time-varying, three-dimensional velocity field data, but is limited to the low speeds typical of liquid helium facilities.

REFERENCES

1. Miles, R. B.; and Nosenchuck, D. M.: Three-Dimensional Quantitative Flow Diagnostics. Lecture Notes in Engineering 45. M. Gad-el-Hak (Editor). Advances in Fluid Mechanics Measurements. Springer-Verlag, Berlin, Heidelberg, 1989.

2. Gad-el-Hak, M.: Review of Flow Visualization Techniques for Unsteady Flows. Flow Visualization IV. Proceedings of the Fourth International Symposium on Flow Visualization, August 26-29, 1986, Ecole Nationale Superieure de Techniques Avdncees, Paris, France. Hemisphere Publishing Corporation.

3. Merzkirch, W.: Flow Visualization. Academic Press, 1974.

4. Trolinger, J. D.: Laser Application in Flow Diagnostics. AGARDograph No. 296, October 1988.

5. Burton, R. A.: A Modified Schlieren Apparatus for Large Areas of Field. Journal of the Optical Society of America, Vol. 39, No. 11, November 1949.

6. Landreth, C. C.; Adrian, R. J.; and Yao, C.-S.: Double Pulsed Particle Image Velocimeter With Directional Resolution For Complex Flows. Experiments in Fluids. Springer-Verlag, Berlin, Heidelberg, 1987.

7. Weinstein, L. M.; Beeler, G. B.; and Lindemann, A. M.: High-Speed Holocinematographic Velocimeter for Studying Turbulent Flow Control Physics. AIAA Shear Flow Contol Conference, March 12-14, 1985, Boulder, Colorado. Paper No. AIAA 85-0526.

8. Weinstein, L. M.; and Beeler, G. B.: Flow Measurements in a Water Tunnel Using a Holocinematographic Velocimeter. AGARD Fluid Dynamics Panel Symposium on Aerodynamic and Hydrodynamic Studies Using Water Facilities, Monterey, California, October 20-23, 1986.

ON THE MEASUREMENT OF SUBSONIC FLOW AROUND AN APPENDED BODY OF REVOLUTION AT CRYOGENIC CONDITIONS IN THE NTF

D. W. Coder
David Taylor Research Center, Bethesda, Md. 20084-5000

S. G. Flechner
J. B. Peterson, Jr.
NASA Langley Research Center, Hampton, Va. 23665-5225

ABSTRACT

After a brief review of fluid mechanics scaling, the rationale for testing in the National Transonic Facility (NTF) for ship hydromechanics purposes is discussed. Some pertinent details of the "1986 Body of Revolution Experiment" are presented along with possibilities for a "Future Body of Revolution Experiment" and a "Future Flat Plate Experiment". Finally NTF testing is considered from the "hydrodynamic" users point of view and a brief comparison is made between the NTF and the Conceptual Helium Tunnel (CHT), which is of major interest to this Conference.

NOMENCLATURE

A	Wake parameter in Fig. 5
A_{MODEL}	Cross-sectional area of model
A_{TUNNEL}	Cross-sectional area of tunnel
c	Speed of sound in fluid
D	Diameter of model
Eu	Euler number $(=p/(\rho V^2))$
F_n	Froude number $(=V/(gL)^{1/2})$
g	Acceleration of gravity
L	Model length
M_n	Mach number $(=V/c)$
n	Power-law exponent in Fig. 4
P	General flow-field point in Fig. 2
P_o	Total pressure of tunnel
p	Local static pressure
q_∞	Dynamic pressure of tunnel
R	Maximum radius of body of revolution model
R_n	Reynolds number based on model length $(=VL/\nu)$
R_θ	Momentum thickness Reynolds number $(=V\theta/\nu)$
r	Radial distance from model centerline
T_o	Total temperature of tunnel
U	Velocity magnitude
V	Free-stream velocity magnitude = ship speed
V_s	Ship speed = free-stream velocity magnitude
We	Weber number $(=\gamma/(\rho V^2 L))$

x	Axial distance from front of bow of body of revolution model
γ	Surface tension
δ	Boundary layer thickness
$\Delta(\ )$	Difference of ( ) which depends on ΔR_n
θ	Boundary layer momentum thickness
λ	Scale ratio: full-scale-to-model
ρ	Density of fluid

INTRODUCTION

Hydromechanics scaling dictates that in order to have exact similitude for model vehicle experiments in water, the nondimensional Reynolds number (R_n), Froude number (F_n), and Euler number (Eu) must be the same as for the full-scale vehicle. These numbers can be derived from the incompressible, constant property version of the Navier-Stokes equation (see White[1]) by appropriate non-dimensionalization. If a free surface (interface between a liquid and gas) such as the ocean surface is part of the physical problem investigated, the Weber number (We) must also be the same at model and full scale for exact similitude. Grossly stated, these nondimensional numbers represent the effects of viscosity, gravity, pressure, and surface tension, respectively, on the dynamics of the fluid.

For a surface ship, the development of the ship boundary layer is mostly governed by the Reynolds number, the development of surface waves is governed by the Froude number, cavitation of the propulsor is governed by the Euler number, spray and bubbles that may occur for breaking waves and interaction between the liquid-gas surface and a model surface are governed by the Weber number, and the wake flow physics probably includes all of the above.

For a deeply submerged body, the flow is not governed by Froude number (free-surface or gravity wave effects) unless the dynamics of the ship needs to be scaled (because of the gravity force). However, the testing of a submerged body in a shallow tow tank requires consideration as to whether the free-surface effects (Froude number effects) must be considered.

For testing of a surface ship or submersible in a wind tunnel, there is the added effect of Mach number (M_n) which has to do with compressibility of the gas. To simulate water, which can be assumed to be an essentially-incompressible liquid for which $M_n \approx 0$ for the problems considered here, it is usually considered sufficient to test in a gas at $M_n \leq 0.3$. This is referred to as low subsonic flow. In the testing of ships in wind tunnels for viscous effects, the usual practice is to represent the surface ship as a double model, the below-surface part of the hull reflected about a flat free surface, or the below-surface part mounted to a flat plate that represents the free surface. If the shape of the free surface, which is a function of Froude number, is known a priori, curved plates may be used to represent the surface for the various Froude numbers. Obviously a submersible can be represented in-toto and is usually mounted in the "potential flow" core of the tunnel using struts and/or cables or on a sting. Because effects of Froude number and Weber number are usually not present in the wind tunnel, the scaling problem reduces to one of obtaining similitude in Reynolds number while keeping the Mach number in the low, subsonic range.

In Table I are given the length, cruse or maximum velocity, and resulting cruse or maximum Reynolds number for a selection of commercial and naval ships in the ocean, using viscosity values for 15 deg C (59 deg F) sea water from Comstock[2]. Notice that the naval ships usually have somewhat higher Reynolds numbers than a comparable-length commercial ship due to their somewhat higher velocities. Notice further that the highest Reynolds numbers of these full-scale ships range from about 100 million to about 4000 million with the aircraft carrier and battleship having the highest Reynolds numbers of all.

Table I   Reynolds numbers for various ship types

SHIP TYPE	NAME	L^{*} (FT)	V^{*} (KTS)	R_n^{**} ($\times 10^6$)
Aircraft Carrier	Nimitz (CVN68)	1040	30	4117
Battleship	Iowa (BB61)	887	35	4097
Passenger Liner	United States	990	28	3658
RO/RO	Westward Venture	734	23	2228
Crude Oil Tanker	Mobil Magnolia	1063	15	2104
Destroyer	Arl. Burke (DDG51)	466	30	1845
Frigate	O. H. Perry (FFG7)	445	29	1703
Navy Auxiliary	Cimarron (AO177)	592	20	1562
Ore Carrier	(not identified)	715	16	1510
BN Submarine	Ohio (SSBN726)	560	20	1478
Amphibious Warfare	(LPD17)	522	21	1447
Mariner Cargo Ship	(various)	523	20	1380
Heavy Lift Transport	Mighty Servant 3	591	14	1092
Victory Cargo Ship	Furman	444	17	996
Attack Submarine	Los Ang. (SSN688)	360	20	950
Liberty Cargo Ship	Jeremiah O'Brien	441	15	873
Clipper Ship	Great Republic	335	14	619
Ferry	John F. Kennedy	285	15	564
Sailing Barque	USCGC Eagle	231	18	549
Mine Warfare	Avenger (MCM1)	224	14	414
SRB Shuttle Ship	Liberty Star	176	14	325
Tug	Valerie F.	138	15	273
Excursion Boat	MV City of S. F.	148	5	98
12-Meter Yacht	Stars & Stripes	47	15	93

* Nominal/estimated values
** $R_n = L(ft) \times V(ft/sec) / \nu(ft^2/sec)$, where $\nu = 1.2791 \times 10^{-5}$ ft^2/sec for sea water at 59°F (15°C) from Comstock (1967)

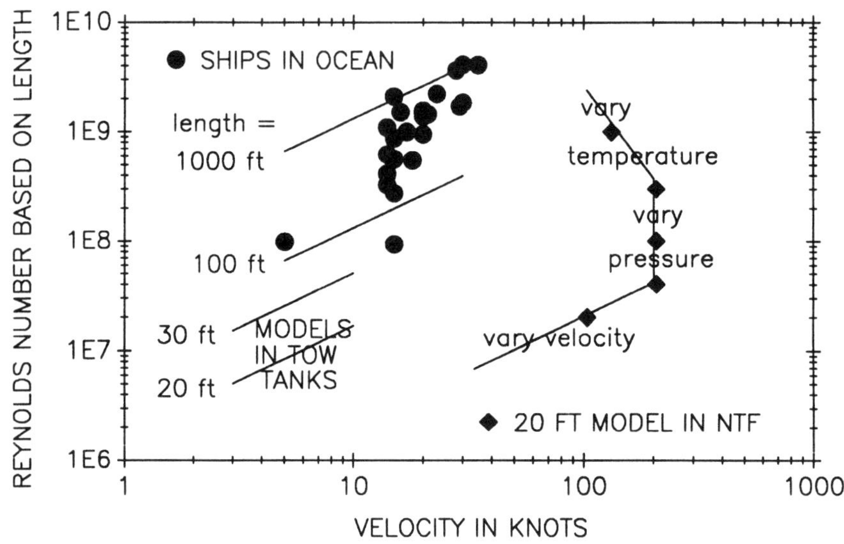

Fig. 1. Reynolds numbers of ships and 20 ft model in NTF.

If the data of Table I are plotted as Reynolds number versus velocity, as in Fig. 1, it is seen that full scale ships tend to occupy the upper left hand quadrant of the figure. Unfortunately, Reynolds numbers obtained when representing ships with 6 to 9 m (20 to 30 ft)-long models (typical size) in typical tow tank facilities (see da Vincent[3]) are around 10 million and these conditions occupy the lower left hand quadrant of the figure. With a large model, 9 m-long (30 ft-long), in the Large Cavitation Channel (LCC), speeds of 30 knots are projected (see Rothblum[4]) and Reynolds numbers of about 100 million might be attainable. There is a possibility that small ships around 30 m (100 ft) in length that have velocities around or below 10 knots might be well represented by a 1/3rd scale model at 3 times the velocity in the LCC. However, in general, it will not be possible to attain maximum Reynolds number similitude even in this facility.

Thus there is a gap in Reynolds number from the typical hydromechanics facilities to full scale that can be filled by testing in the National Transonic Facility (NTF), a cryogenic wind tunnel located at NASA/Langley in Hampton, Virginia (see McKinney and Baals[5]). The capability to span the Reynolds number range of interest, about 10 million to 1000 million, is shown on the right side of Fig. 1. The NTF has a 6.1 m-long (20 ft-long) test section which has a square cross-section 2.5 m (8.2 ft) on a side with flat corner fillets. The working fluid is air when operating "warm" (at room temperature) and gaseous nitrogen when operating "cold" (at cryogenic temperatures). The low temperatures are obtained in the tunnel by atomizing liquid nitrogen (liquefaction temperature of -196 deg C (-320 deg F) at atmospheric pressure) into the gaseous nitrogen ahead of the impeller fan. The tunnel can be pressurized to about 9 atmospheres. Both low temperature and high pressure allow attainment of high Reynolds numbers according to the following equation derived using the ideal gas law, the definition of Mach number, and the proportional relationship between viscosity and the power (0.9 for air or nitrogen) of absolute temperature[5]:

$$R_n \varpropto LM_nPT^{-1.4}$$

Thus to obtain the highest Reynolds number (R_n), it is desirable to have the largest model length (L) possible (one that will fit in the test section and not cause prohibitive blockage), test up to the highest allowable Mach number while maintaining low subsonic flow conditions to simulate water flow (M_n up to about 0.3), obtain the highest pressure (P) possible, and obtain the lowest temperature (T) possible. With a model length of about 6.1 m (20 ft), the length of the test section, it is possible to obtain Reynolds numbers from about 10 million to above 1000 million (solid line on the right side of Fig. 1).

REQUIREMENTS FOR HIGH REYNOLDS NUMBER TESTS IN THE NTF

The Reynolds number capability of the NTF can be used to perform experiments for several different purposes of importance to ship hydromechanics:

(a) <u>Develop scaling laws</u>: The range of Reynolds number allows the determination of the Reynolds number effect from typical hydrodynamic facility values to full scale. After this effect is determined, data from other model experiments including much historical data may be extrapolated to obtain full scale predictions. For example, the "nominal wake", stern flow without a propulsor, for a ship shape may be measured on a small model (about 20 ft long) in a tow tank or low speed atmospheric wind tunnel and then extrapolated to full scale in order to be used as the nominal wake that is needed to start the full scale propulsor design.

(b) <u>Evaluate and improve model scale experiments</u>: Several model experiments are used, hopefully, to predict directly the full scale behavior. In order for these experiments to work properly, the flow dynamics needs to be scaled or at least represent full scale behavior in its overall integrated effect. An example of this would be scale model experiments to obtain safe maneuvering conditions by using a freely running model that is maneuvered in a basin. For small models which are operated at Froude scaled velocities, in order to scale gravity effects, and in the same fluid as full scale, the Reynolds number is $\lambda^{3/2}$ times smaller than full scale (λ = geometric scale factor). Thus hydrodynamic characteristics that are very much a function of Reynolds number, such as control surface stall, have to be controlled in order to approach a scaled maneuver. Testing of control devices, such as boundary layer tripping devices, over the Reynolds number range might point the way toward helping to reduce the effect of flow separation on control surfaces at low Reynolds number and extend the range of maneuvering similitude for the model.

(c) <u>Evaluate full scale systems</u>: A third experimental purpose is to test a full scale device at full scale Reynolds number conditions. This is essentially an evaluation experiment that would demonstrate the relative performance of various competing designs. For example, fillets on control surfaces tend to reduce downstream flow variation for the inflow to a propulsor (see Kubendran and Harvey[6]). It is very important to "clean up" this flow to improve cavitation performance and to reduce vibrations of the propulsor.

(d) Provide data for CFD validation: At the present time, the flow around a full scale ship with all its appendages cannot be accurately calculated. However great strides have been made and are being made at the present time. The computational fluid dynamics (CFD) community needs data over a range of Reynolds numbers to guide the analytical development and to validate the emerging codes. This may be the direction that NASA is taking with the "Numerical Wind Tunnel" concept[7] -- that wind tunnels will be used for validating computer codes and the computer will be used to perform interpolation and refinement of the geometry effects instead of building many variant models. The ship hydromechanics community may be taking this direction also - - conducting CFD validation experiments in tow tanks and wind tunnels rather than performing evaluation experiments with many models (see Morgan and Lin[8]). Perhaps eventually, much of the ship hydromechanics evaluation work can be performed with "numerical towing tanks".

In 1986 an experiment with a 6.1 m-long (20 ft-long) appended body of revolution was performed in the NTF that had as its purposes items (a) and (d) above. Future experiments with a body of revolution and a large flat plate in the NTF are envisioned that would be multi-purpose, (a) through (d) above, and would considerably extend our knowledge of the Reynolds number effects on the hydromechanics of ships. The previous and envisioned future experiments are briefly discussed in the next three sections.

1986 BODY OF REVOLUTION EXPERIMENT IN THE NTF

The model for the 1986 experiment was a 0.56 m-diameter (22 in.-diameter), 6.1 m-long (20 ft-long) appended body of revolution that was developed for an earlier experiment in the NASA/Ames Twelve Foot Pressure wind tunnel (TFP) (see NASA[9] and Coder[10]). The model was determined to be hydraulically smooth for all Reynolds numbers tested and model boundary layers were tripped with a 0.25 mm-diameter (0.010 in.-diameter) wire at 5 percent of the length and at 5 percent chord of all appendages. The model was adapted for use in the NTF and a new mounting system was built to take care of the larger tunnel dynamic pressure (design pressure up to 62 kPa (1300 psf)) and the low temperature (design temperature down to -196 deg C (-320 deg F)). A schematic of the model mounted in the NTF is given in Fig. 2. The model was supported by a large forward strut and smaller cruciform struts that were attached to the tips of the stern appendages. The NTF ceiling and floor slots were covered for this experiment.

The model had pressure taps on its crown line (top of the model) at every 1 percent of model length up to the stern appendages, several circumferential rings of taps near the front of the model and at three axial positions on the parallel midbody, and several taps on the hull in between the stern appendages. Other instrumentation included three total head tube boundary layer rakes that were located at three axial locations on the parallel midbody and a special

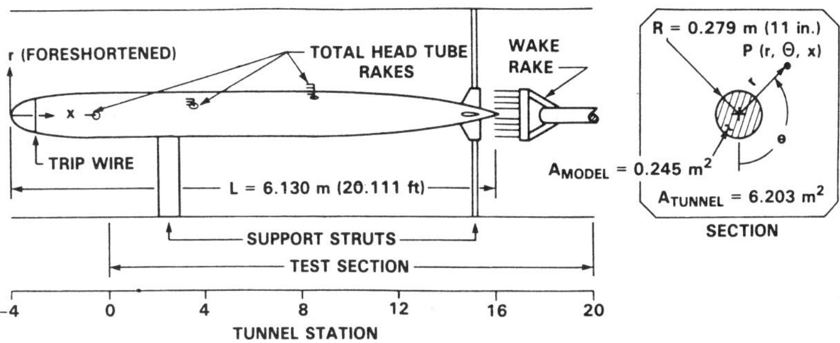

Fig. 2. Schematic of model in test section of NTF.

purpose wake rake that holds 20 probes (10 five hole pitot tubes and 10 tri-axial hot film probes) with their measurement volumes in a vertical plane at the stern of the model (see closeup view of probes in Fig. 3). Since there is differential structural expansion between the test section walls and the tunnel arc sector strut located in the diffuser of the NTF, a platform was built from the test section walls to support the roll bearing for the wake rake sting so the location of the bearing could be kept constant relative to the model during cryogenic operation. The wake rake sting attached to the arc sector strut roll mechanism with a spline connection which allowed relative axial motion and still allowed use of the mechanism to roll the wake rake. The Reynolds numbers obtained were from about 20 million to about 1000 million for the test conditions shown in Table II.

Table II Test conditions in the NTF

R_n (x10^6)	M_n	P_o (psi/kPa)	T_o (°F/°C)	q_∞ (psf/kPa)	Turbulence Intensity[*]	Solid Blockage
21	0.156	15.5/107	100/38	37.5/1.80	0.07%	3.9%
44	0.312	15.4/106	"	141/6.75	0.13%	"
102	0.312	38.6/266	"	354/16.9	0.18%	"
308	0.312	117/807	"	1075/51.5	0.60%	"
1046	0.312	117/807	-230/-146	1075/51.5	0.60%	"

[*]Nominal values excerpted from work by W.B. Igoe on a doctoral dissertation to be submitted to the George Washington University.

The instrumentation performance in the high dynamic pressure and low temperature of the NTF was mixed. The static pressure and total head tube pressure measurements were accomplished by connecting the backs of the pressure taps and tubes to the pressure module inside the model with flexible tubing. These modules were electrically heated during cryogenic tunnel conditions. One out of the four modules froze so that some of the

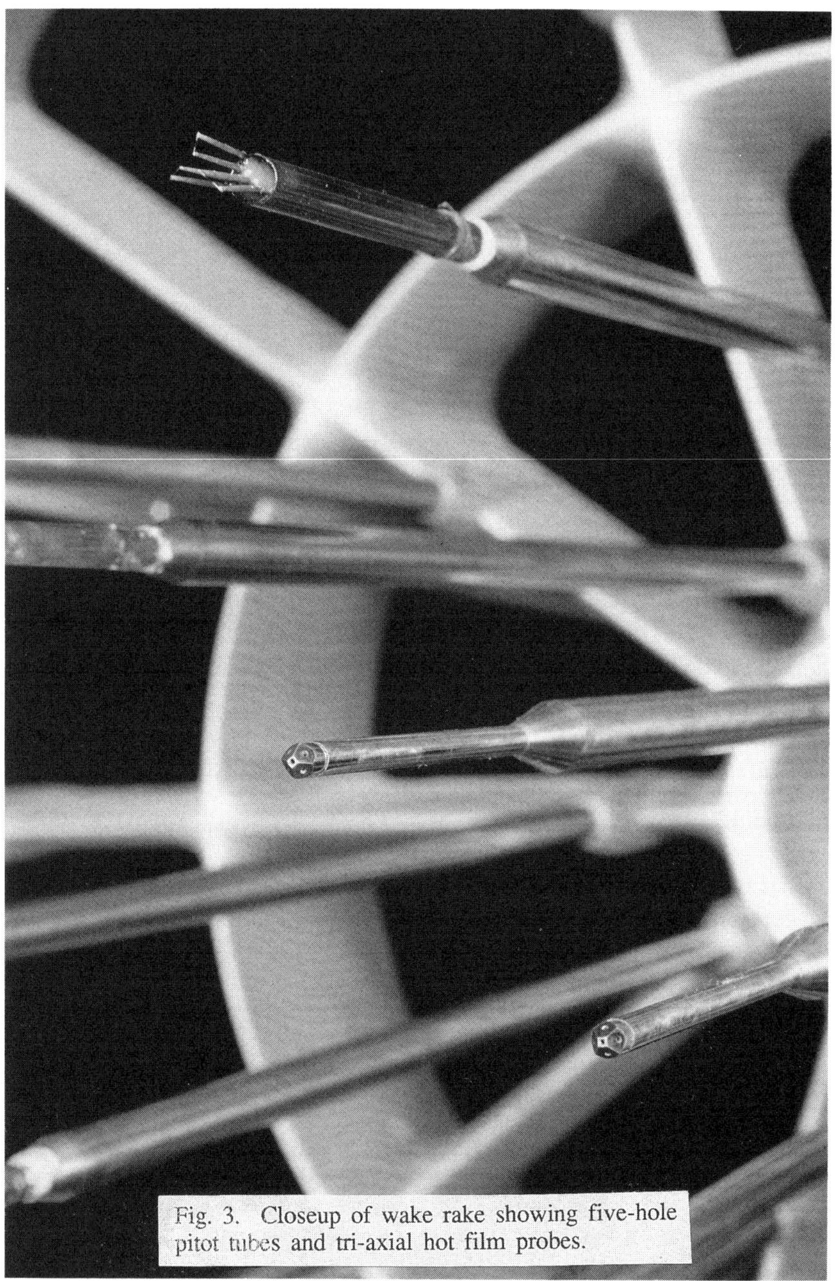

Fig. 3.   Closeup of wake rake showing five-hole pitot tubes and tri-axial hot film probes.

hull pressure data for the cryogenic condition were lost. The three total head boundary layer rakes and their pressure modules worked satisfactorily for all tunnel conditions.

The data obtained from the boundary layer rakes (see Coder[11]) are unique as far as the high Reynolds number is concerned and were analyzed together with data from the earlier, lower Reynolds number TFP experiments (see Coder[12]) with the same model. Some of the results of the analysis taken from Coder and Rubin[13] are shown in Figs. 4 and 5. Here it is seen that the power law exponent (n), the boundary layer thickness (δ), and the boundary layer wake parameter (A) (see White[1] for definitions) continue to be a function of Reynolds number even at the highest Reynolds numbers achieved.

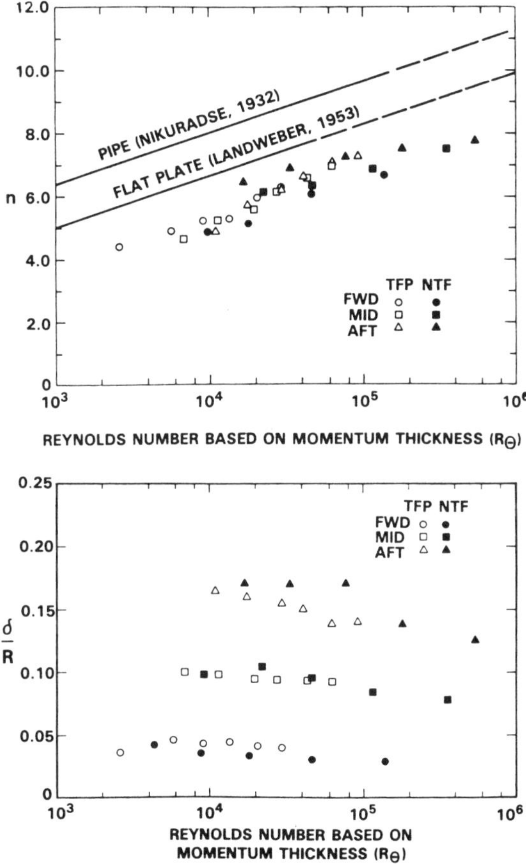

Fig. 4. Effect of Reynolds number on power-law exponent and boundary layer thickness.

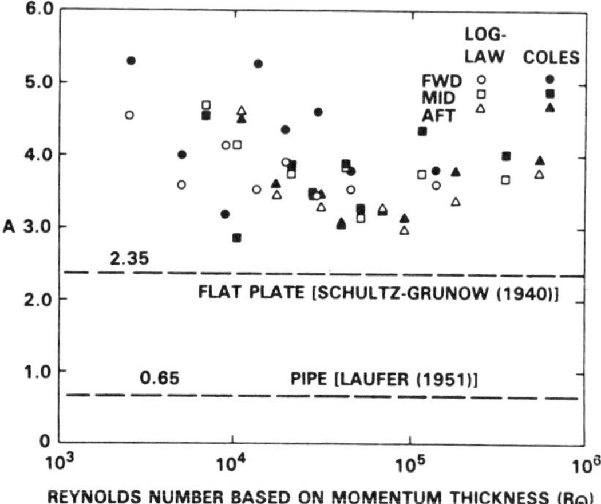

Fig. 5. Effect of Reynolds number on boundary layer wake parameter.

The wake rake was calibrated in the open tunnel. The wake rake sting was set at various pitch angles and the rake was rolled through 360 deg. During these calibrations it was observed that some "frosting" occurred on the probes during cryogenic conditions and some of the hot film probes were damaged. It is speculated that the hot film damage was due to non-atomized drops of liquid nitrogen that were impacting and breaking the films. By reducing the highest Reynolds number from about 1500 million to about 1000 million, the damage problem was significantly mitigated. This was then set as the maximum test Reynolds number. The frosting of the hot film probes was eliminated, it was thought, by simply turning the probes on (the frost was observed via video camera to disappear as soon as the current was turned on). During the calibration, the frost on the pressure probes may have caused unrealistic pressures due to frost piling up near the front stagnation point and clogging the hole and/or changing the local geometry of the probe head. This condition was remedied by operational means -- the temperature in the tunnel was lowered in steps and the tunnel gas that had moisture in it (from the out gassing of small amounts of water in the tunnel insulation) was purged and dry gas was introduced. Using this method, the pitot tube and hot film probe measurements appeared to be satisfactory during the testing with the model in place. A more detailed description of the moisture problem and the correction to the problem may be found in Gloss and Bruce[14].

The components of velocity in the stern measuring plane were obtained with the pitot tubes. Some of the analyzed results from Coder[11] are shown in Fig. 6. Here the incremental change in velocity (change in profile fullness) as a function of the incremental change in Reynolds number is shown to be quite different from that which is predicted by axisymmetric boundary layer calculations. This is a significant finding which indicates that the current extrapolation techniques used to predict the full scale propulsor inflow from typical tow tank Reynolds number data need improvement.

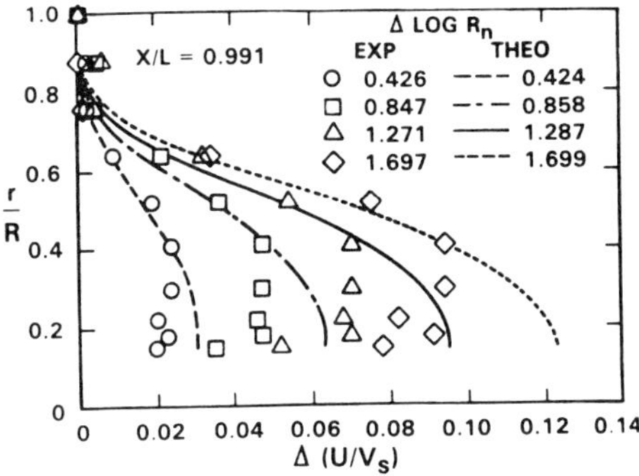

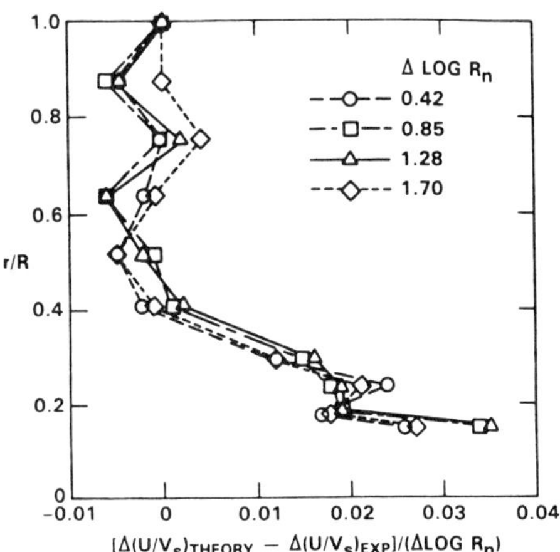

Fig. 6. Effect of Reynolds number on nominal wake velocity profiles.

FUTURE BODY OF REVOLUTION EXPERIMENT IN THE NTF

Many more measurements than were made for the 1986 experiment are envisioned for possible future experiments in the NTF. In order to accomplish these measurements, the model capabilities will have to be expanded. The major model modifications include:

(a) modification to the model mounting system to allow remote pitch and yaw attitude changes,

(b) modification to the appendages to allow them to be remotely actuated throughout their normal angular range,

(c) installation of a motor inside the model to allow the operation of model propulsors,

(d) modification to the wake rake to allow remote positioning to various positions downstream of the model, and

(e) installation of improved and new instrumentation to improve the measurements previously made and to expand measurement capability.

The expanded model capabilities will allow the following measurements to be made:

(a) forces and moments on the model, appendages, and propulsor components,

(b) static pressure distribution on the hull, appendages, and propulsor components,

(c) boundary layer and wake steady velocity components, turbulence (intermittency and single point velocity fluctuation correlations), and static pressure,

(d) hull separation patterns and velocities,

(e) appendage/hull junction flow velocities and pressures,

(f) appendage tip vortex flow velocities and pressure,

(g) flow around, through, and downstream of propulsors.

FUTURE FLAT PLATE EXPERIMENT IN THE NTF

The NASA Langley Research Center has been developing a fundamental experiment concerning high Reynolds number turbulent boundary layers on a flat plate (see Saric and Peterson[15]). In this regard the ship hydromechanics community has a similar interest in high Reynolds number turbulent boundary layer research but with a slightly different focus. Whereas NASA is interested across the range of Mach number from subsonic to supersonic, the ship hydromechanics community is mostly interested in the low subsonic flow regime ($M_n \leq 0.3$). Also there is extreme interest in turbulent boundary layer velocity and pressure fluctuations because of the noise applications. Because of these two focuses, the flat plate work should include the following features:

(a) The length Reynolds number for undisturbed (by the wall) flow should reach at least 1000 million.

(b) The Mach number should be 0.3 or below.

(c) Static pressure and temperature should be measured.

(d) Wall shear stress and plate vibration should be measured.

(e) Boundary layer transition location and profiles should be measured.

(f) TBL fluctuating velocities and pressures should be measured.
(g) In the future, other basic conditions and geometries including various pressure gradient effects on flat plate flow, forward and rear steps, and cavity flows should be investigated.

CONSIDERATIONS OF THE USER AND HOW THEY MAY RELATE TO THE PROPOSED LIQUID HELIUM TUNNEL

As can be seen by what has gone before, the NTF somewhat satisfies the high Reynolds number need for laboratory controlled experiments concerning the viscous flow over ships. While there is some desire to obtain Reynolds numbers of about 4000 million (see Fig. 1), values of around 1000 million are considered adequate for most purposes. It is also important to obtain the range from typical hydromechanics facilities of about 10 million to the full scale values of about 1000 million in order to obtain scaling information for extrapolation purposes. In order to obtain the range and highest Reynolds number, the whole length of the test section has been occupied and the blockage has been allowed to attain 3.9 percent (at mid model). For the Future Body of Revolution Experiment it would be desirable to pitch and yaw the model to angles of 15 deg or more but the increase in blockage and proximity to the tunnel walls limit practical testing in the NTF to about 5 deg for a 6.1 m (20 ft) model. For the ship situation, where the wake, which flows into the propulsor, changes significantly with angle, 5 deg is sufficient to characterize this effect for normal mild maneuvers. However, it would also be desirable to test up to 15 deg for extreme maneuvers. In order to do this in the NTF, a smaller model could be used but this would mean less-than-full-scale Reynolds numbers.

Another consideration when testing a large model system in a wind tunnel is the model mounting system and its interference effects on the flow field. The model mounting system shown in Fig. 1 utilizes a forward strut and struts attached to the tips of the stern appendages. This was done to minimize the effect of the struts on the hull boundary layer. Previously, Coder[12] used a second strut near the stern to support the body when it had no stern appendages. These struts obviously have an interference effect on the flow and even the struts attached to the end of the appendages have some effect (see Mellish and Coder[16]). Although Mellish and Coder[16] show that the interference effects of a mounting strut attached to the end of an appendage might be mitigated by suction and blowing near the appendage/strut intersection, this idea is promoted only for models that are to be tested with zero deg on appendages and struts. The strut interference effect is much more severe for maneuvering conditions where the struts can develop lift (and circulation) themselves and can interfere with the lift (and circulation) and tip vortices generated by the appendages. One interesting idea that would eliminate the strut interference effect has been promoted by NASA -- the Magnetic Suspension and Balance System (MSBS), see Lawing et al.[17] The idea is to suspend the model with magnets and measure the fluid dynamic loads through measurement of the magnet currents that are necessary to hold the model in place. For a large, heavy model, such as the

6.1 m-long (20 ft-long) body of revolution model in the NTF, which is exposed to large dynamic pressures in the tunnel, the MSBS may be impractical. However, for smaller models that might be used in the proposed liquid helium tunnel, this may be quite practical.

The characteristics of the proposed liquid helium tunnel may be found in Donnelly et al.[18] and Smith and elsewhere in the <u>Proceedings of the Seventh Conference on Low Temperature Physics</u>. A summary of the advantages and disadvantages of the proposed helium tunnel, now referred to as the "Conceptual Helium Tunnel" (CHT), as compared to the NTF are now considered briefly. This summary represents an expansion of material that the first author presented at the Conference and benefits from information gleaned from conference participation. In some aspects, this summary represents the opinion of the authors and does not necessarily represent the official attitude of the Navy or NASA. The summary is not meant to be inclusive of all the factors that may come into play when deciding whether or not to build the Conceptual Helium Tunnel but rather a starting point for further discussion. The summary will consider the comparison under the topics of Test Conditions, Tunnel Characteristics, Model Characteristics, and Measurements.

TEST CONDITIONS

The attainable Reynolds number range for the NTF and a 6.1 m-long (20 ft-long) model has been demonstrated to be from about 20 million to 1000 million. The Reynolds number for the CHT shown in Table III would be up to about 300 million by using He I and up to 1000 million by using He II. Presumably Reynolds numbers of about 10 million can be obtained also. Whereas the Mach number for the NTF is between about 0.15 and about 0.3, the Mach number would always be low (≤ 0.1) for the CHT. The fluids used in the two tunnels are air (around room temperatures) and gaseous nitrogen (cold temperatures) for the NTF and liquid HeI (very cold temperatures) and liquid HeII (near absolute zero) for the CHT. Air, Nitrogen, and HeI are Newtonian or "Navier-Stokes" fluids but HeII is a "super" fluid.[19] It is left to demonstrate that HeII, a super fluid, can be used to simulate the flow of water, a Navier-Stokes fluid, in all important regards. The CHT would have the capability to test with both HeI and HeII at Reynolds numbers up to 100 million. If the results at this lower Reynolds number of 100 million for the two fluids, one Newtonian and the other a superfluid, showed no differences, then it could be argued that the results for the superfluid at the higher Reynolds numbers of 1000 million would be representative of Newtonian fluid results. Whereas the NTF can operate for a few hours at the highest Reynolds number until it runs out of liquid nitrogen in the storage tanks, the CHT is projected to operate continuously.

Table III   Comparison of test conditions between NTF and CHT

Item	NTF[*]	CHT[**]
Reynolds Number	$10^7 \leq R_n \leq 3 \times 10^8$ with air 10^9 with Nitrogen	$R_n \leq 3 \times 10^8$ with HeI $R_n \leq 10^9$ with HeII
Mach number	$0.15 \leq M_n \leq 0.3$	$M_n \leq 0.1$
Kind of fluid	Gases: air = Newtonian fluid N_2 = Newtonian fluid	Liquids: HeI = Newtonian fluid HeII = superfluid
Time on condition	Continuous with air 1/2 to 6 hr with N_2	Continuous for both Intermittent for even higher R_n

[*]National Transonic Facility with a 6.1 m-long (20 ft-long) body of revolution model.
[**]Conceptual Helium Tunnel with a 0.5 m-long (1.5 ft-long) body of revolution model.

TUNNEL CHARACTERISTICS

The linear dimensions of the CHT as shown in Table IV would be about 8 times smaller than the NTF, 0.3 m square test section instead of a 2.5 m square test section, which means about 500 times smaller in volume. This means that the CHT would be much less expensive to build even though the CHT requires an expensive refrigeration plant along with the tunnel. Operational costs involve a large staff, electrical power for the tunnel, and nitrogen consumption for the NTF versus a medium staff, modest power for the tunnel, and electrical power to run a refrigeration plant. The large pressure, low temperature, and nitrogen hazard make the NTF more dangerous to operate than the low pressure, low temperature, and Helium for the CHT. Access to the test section appears to be a considerable problem for the CHT. Although the NTF requires up to 4-5 hrs to cool down or warm up, the CHT has to have a thermal jacket around the tunnel which would have to be disassembled to get to the test section. One idea which has been advanced to reduce the model access time in the CHT is to provide a "model lock" so that the model could be passed through the jacket and into the test section. If a MSBS were included in the tunnel, the model might be able to "fly" into the test section using the MSBS.

MODEL CHARACTERISTICS

As shown in Table V, models for the CHT could be on the order of 0.5 m (1.5 ft) long and would cause about the same blockage as for the NTF. Even though smaller models usually mean they are less costly, the model surface finish to obtain a hydraulically smooth body would be very difficult to obtain.

Table IV   Comparison of tunnel characteristics between NTF and CHT

Item	NTF[*]	CHT[**]
Test section size	2.5mx2.5m crossection 6.1m-long	0.3mx0.3m crossection 1.2m-long
Overall tunnel size and capital costs	Relatively large and expensive	Relatively small and of medium expense
Cooling technology and operating costs	Atomize liquid N into gaseous N_2 Expense is large staff and N_2	Cool recirculating liquid He with refrigeration plant Expense is power to run plant
Pressure	1 to 9 atm possible	1 atm
Temperature	38°C (100°F) to -196°C (-320°F) possible	Down to below 2.172°K for HeII
Tunnel dynamic pressure	Up to 51.5 kPa (1075psf) for BOR tests	Very low
Access	Isolation valves for test section Up to 4-5 hrs each to cool down & warm up	Whole tunnel in jacket access very limited Pass-through model lock
Safety	N_2 hazardous Pressure high Dynamic pressure high Temperature low	He innocuous Pressure low Dynamic pressure low Temperature very low

[*]National Transonic Facility with a 6.1 m-long (20 ft-long) body of revolution model.
[**]Conceptual Helium Tunnel with a 0.5 m-long (1.5 ft-long) body of revolution model.

Table V   Comparison of model characteristics between NTF and CHT

Item	NTF[*]	CHT[**]
Length	6.1 m (20 ft)	0.5 m (1.5 ft)
Diameter	0.55 m (1.8 ft)	0.05 m (0.15 ft)
Blockage in tunnel	3.9%	2.2%
Surface finish for admissible roughness	0.61 micron (24 μin.)	0.05 micron (1.8 μin.)
Material	Aluminum	Aluminum and ferromagnetic for MSBS
Strength	Designed to high tunnel q_∞ 62 kPa (1300 psf)	Designed to low tunnel q_∞
Weight	Heavy especially with propulsion and actuators/balances	Fairly light
Internal room for Instrumentation	Sufficient room for instrumentation, heaters, actuators, etc.	Very little room need telemetering and heater dev. with MSBS

[*]National Transonic Facility with a 6.1 m-long (20 ft-long) body of revolution model.
[**]Conceptual Helium Tunnel with a 0.5 m-long (1.5 ft-long) body of revolution model.

Also in such a small model there is not much room to include instrumentation inside, especially since there is a need for heating conventional instrumentation.

MEASUREMENTS

Because of the small size of the model, it would be very difficult to measure the flow field around the model with conventional instrumentation. The resolution volume for instrumentation for the NTF is just about acceptable and the CHT model would be about an order of magnitude smaller (see Table VI). About all that could be measured would be the gross flow features. What is more possible is the measurement of overall forces, especially with a MSBS so that strut interference effects can be eliminated. It may also be possible with a MSBS in the CHT to measure overall model forces during small-angle (less than about 5 deg) maneuvers.

Table VI   Comparison of measurements between NTF and CHT

Item	NTF[*]	CHT[**]
Measurement volume and uncertainty	Suitable for for Ship Hydro.	Not suitable with current instrumentation
LDV	Needs development of N_2 cooled laser or fiber optic system	Lower temp makes probs more difficult Seeding a problem
Forces	Room for balances and dynamometers	Not much room but some might be possible
MSBS	Difficult due to weight and loading Not too promising Cold available for superconducting magnets	Demonstrated feasible Cold available for superconducting magnets

[*]National Transonic Facility with a 6.1 m-long (20 ft-long) body of revolution model.
[**]Conceptual Helium Tunnel with a 0.5 m-long (1.5 ft-long) body of revolution model.

SUMMARY

Much of the need for testing ship models at high Reynolds numbers and over a range of Reynolds numbers can be satisfied by the characteristics of the NTF. Using a 6.1 m-long (20 ft-long) body of revolution model and conventional instrumentation, measurements of boundary layer flow, pressure distributions, and wake flow have been made for Reynolds numbers from 20 million to 1000 million and show that the flow field is significantly affected by Reynolds number. Future experiments in the NTF are envisioned with model propulsors installed, with actuators to remotely change the model attitude, the deflection angle of appendages, and the longitudinal location of the wake rake, and force and moment balances and dynamometers. Limitations in using the National Transonic Facility (NTF) with this large, heavy model are: (a) unresolved strut interference effects and (b) the inability of obtaining model angles of attack greater than about 5 deg due to blockage problems.

It is projected that the Conceptual Helium Tunnel (CHT) would be capable of attaining Reynolds numbers of 1000 million with the use of HeII which is a superfluid. There is concern that a superfluid would not represent all of the desired characteristics of water at high Reynolds numbers. However, the CHT would allow testing of both HeI, a Navier-Stokes fluid, and HeII, a super fluid, at the same Reynolds number and Mach number. This would go a long way toward determining the nature of HeII. Since the CHT is small, it is unlikely more than gross features of the flow field can be measured with conventional instrumentation. The likelihood of measuring forces and moments is good and strut interference effects can be eliminated using a MSBS. Although access to the tunnel would not be good, the idea of using a model-lock would reduce this problem for some applications. Although only small angles of the model could be tested in the CHT (as in the NTF), it may be possible to perform some unsteady maneuvers in the CHT with a MSBS.

ACKNOWLEDGEMENTS

The experiment that is discussed in the text, the 1986 Body of Revolution Experiment in the NTF, was jointly sponsored by the David Taylor Research Center (DTRC) Internal Research (IR) program, the Office of Naval Research (ONR) Fluid Dynamics program, and the NASA/Langley Research Center (LaRC) High Reynolds Number Research program. The initial model development and earlier testing in the TFP was sponsored by the Naval Sea Systems Command (NAVSEA) under the Submarine Drag Reduction Program (SDRP). The planning for the future possible experiments in the NTF and the preparation of this paper were jointly funded by the DTRC IR program and the DARPA Advanced Submarine Technology (AST) program.

REFERENCES

1. F. M. White, Viscous Fluid Flow (McGraw-Hill, N. Y., 1974)
2. J. D. Comstock, Principles of Naval Architecture (SNAME, 1967)
3. M. C. da Vincent, NSRDC Report 3039 (1970)
4. R. S. Rothblum, NATO Defence Research Seminar on Advanced Hydrodynamic Facilities (NATO, 1982).
5. L. W. McKinney and D. D. Baals, NASA CP 2183 (1981).
6. L. R. Kubendran and W. D. Harvey, 3rd Aerodyn. Conf. (AIAA, 1985).
7. "Numerical Aerodynamic Simulation Program Plan", NP-1000-02-C00 (NASA Ames Research Center, 1988).
8. W. B. Morgan and W. C. Lin, Proc. of the 18th ITTC (1987).
9. "Ames Research Facilities Summary", (NASA Ames Research Center,1974).
10. D. W. Coder, NASA CP 2319 (1983).
11. D. W. Coder, Proc. of the 17th Symp. on Naval Hydro. (ONR, 1988).
12. D. W. Coder, Proc. of the 14th Symp. on Naval Hydro. (ONR, 1982).
13. D. W. Coder and M. B. Rubin, Proc. of the Symp. on Hydro. Perf. for Enhancement of Marine Applications (ASME and SNAME, 1988).
14. W. E. Bruce, Jr. and B.B. Gloss, Special Course in Advances in Cryogenic Wind Tunnel Technology (von Karman Institute for Fluid Dynamics, 1989).
15. W. S. Saric and J. B. Peterson, 13th Aerodyn. Testing Conf. (AIAA, 1984).
16. R. W. Mellish and D. W. Coder, DTRC Report SHD-1129-01 (1988).
17. P. L. Lawing, R. A. Kilgore, and D. D. Dress, Aerospace America (March 1989), p. 34.
18. R. Donnelly, M. Smith, and C. Swanson, Physics World (February 1990).
19. R.J. Donnelly, Scientific American (November 1988), p.100.

WATER TUNNELS

Lisa J. Bjarke
NASA Ames-Dryden Flight Research Facility, Edwards, CA 93523-5000

ABSTRACT

This paper provides background on some of the uses of water tunnels in aerodynamic and hydrodynamic testing. The NASA Ames-Dryden Flow Visualization Facility is described in detail and its capabilities are discussed. Examples of previous tests typical of this facility are given.

INTRODUCTION

When dealing with any kind of aerodynamic or hydrodynamic vehicle, it is extremely helpful to understand what is happening around its flow field around this vehicle. Historically, water tunnels have provided valuable insight into fundamental fluid mechanics of two- and three-dimensional configurations. Water tunnels are cheaper to use and less difficult to schedule than are wind tunnels and/or flight test; thus they can be used to accomplish flow visualization in a time- and cost-effective manner.[1]

As designs of aerodynamic and hydrodynamic vehicles become more complex, water tunnel utilization has increased. There are several different types of facilities in use around the world,[2] however, this paper will focus only on the NASA Ames-Dryden Flow Visualization Facility. This in-house facility was built in 1982 for approximately $65,000. Its purpose is to provide an environment for conducting flow visualization studies on research aircraft. A description of the facility, a discussion of its capabilities, and examples of previous tests are given in this paper.

SPECIFICATIONS

The Flow Visualization Facility is a continuous flow, closed return water tunnel. Figure 1 is a photograph of the facility and a schematic is shown in figure 2. The facility has a vertical test section in which the water flows downward. The cross-sectional dimensions of the test section are 16 in. by 24 in. and it is 72 in. long.

Fig. 1.   Overall view of facility.

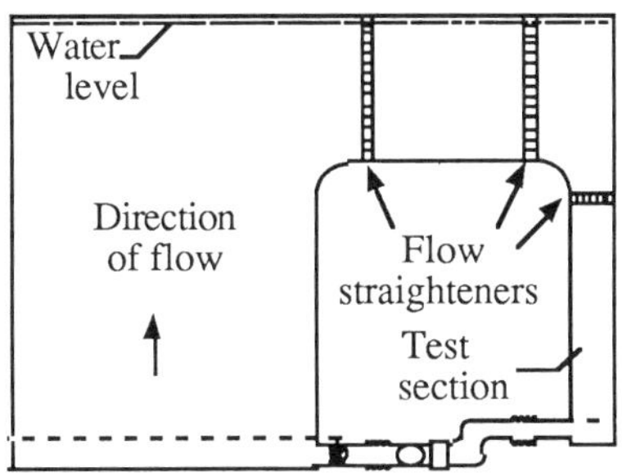

Fig. 2.   Schematic of the Flow Visualization Facility

126   Lisa J. Bjarke

Access to the interior of the test section for model installation or model changes is provided by a 16-in diameter door located on the side of the test section (figure 3).

Flow is provided by a 25-horsepower motor driving a centrifugal pump. The motor is equipped with a speed control which can vary the flow rate from 0.5 in./sec to 18 in./sec.

Fig. 3.   Facility test section.

Figure 4 shows the Reynolds number as a function of flow velocity. This water tunnel is strictly a low Reynolds number facility. However, it has been found that certain types of flow phenomena, namely vortex and highly separated flows, do not necessarily depend on Reynolds number, and therefore can be studied in a water tunnel.[2] Figure 5 shows the turbulence level measured in the center of the test section as a function of flow velocity. These measurements were made using a hot-film anemometer.

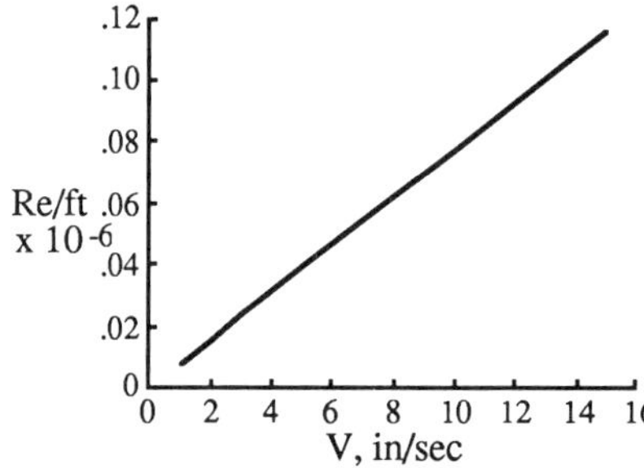

Fig. 4.   Unit Reynolds number range at 68 °F.

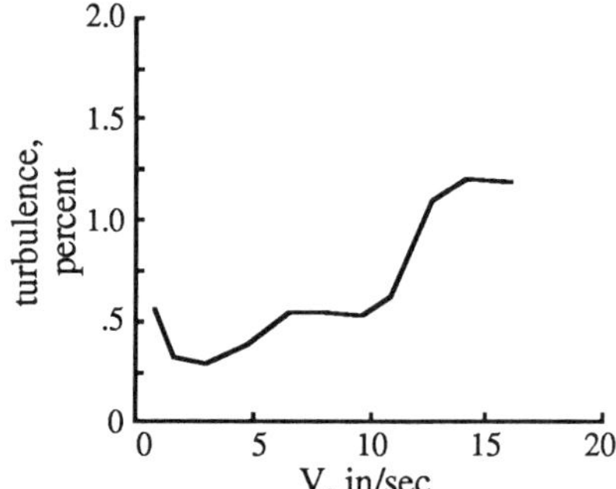

Fig. 5.   Centerline turbulence.

CAPABILITIES

Data Acquisition

Since the majority of the tests performed in the Flow Visualization Facility are visual in nature, it follows that the primary method of data acquisition is the use of photo/video documentation equipment. To provide high-quality documentation at a particular instant in time, still photographs are taken. Two 35-mm cameras are available as well as a 120-type camera. To document the motion of the flow field, two video cameras are used. A screen splitter is available so that the planform and side view can be recorded simultaneously. The video system also includes a text generator which can create title screens and display test conditions.

On occasion, a laser doppler velocimeter has been used to obtain quantitative data in this facility.[3] However this system was provided by the user and at present NASA Ames-Dryden does not own such a system.

Propulsion System Simulation
     Past experience has shown that is important to properly model the flow through aircraft inlets to simulate mass flow through an engine. This is typically accomplished by incorporating flow through inlets in the model and ducting the water from the model to the exterior of the test section. The flow rate can be controlled from outside the test section. Similarly, an engine exhaust jet is modeled by ducting water from an external pressurized water source to the model.

Motion Control System
     The test section is equipped with a model support that can automatically position the model. It consists of a motion controller, servos, joystick and model support arm. A computer is used to run both the motion control system and the character generator for the video camera system described previously. The model can be positioned in angle of attack and sideslip. The model can be positioned in one of two ways: 1) the joystick can be used to position the model and that position is fed back through the computer to the character generator, or 2) the desired angular position can be input by the user by way of the keyboard and the position is then fed to both the model controller and the character generator. Both static and dynamic positioning of the model is possible. For a typical model (18 in. long) the limits on angle of attack are from -30° to 90° and on angle of sideslip from -10° to 15°. Figure 6 shows a simple schematic of the model support and servos used to position the model.

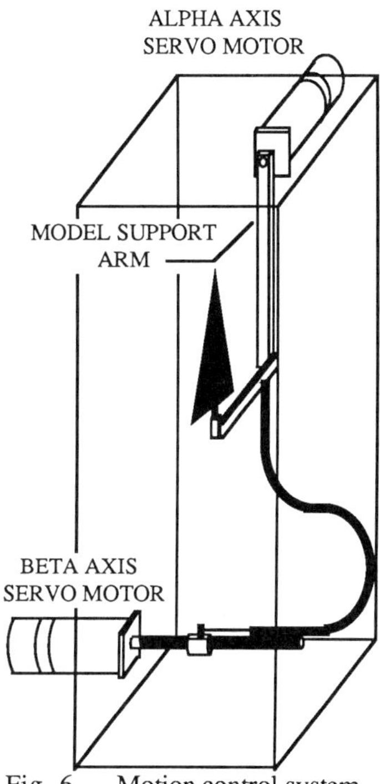

Fig. 6.   Motion control system.

FLOW VISUALIZATION TECHNIQUES

Dye Injection
     The most common flow visualization technique used in the Flow Visualization Facility is that of dye injection. Typically, a mixture of vegetable food coloring and water is injected into the water from ports on the model's surface and, acting as a tracer, the dye is entrained into the flow field. Twelve pressurized dye reservoirs, each controlled by their own valve, provide the model with different colors of dye. The user has a choice of six different colors. Several of the model's dye lines can be manifolded

128   Lisa J. Bjarke

to a single dye reservoir for one-color use. An external dye probe is also available to
visualize the flow away from the model's surface.

Laser Light Sheet
      The laser light sheet technique is very useful to visualize the cross section of a
flow field. Ether fluorescent dyes or a particulate such as aluminium powder is
introduced into the flow field to function as tracer material. A very thin sheet of laser
light is used to illuminate the tracer material in the flow field. The fluorescing dye or
the reflecting particles provide the pattern of the flow field cross section. A four-watt
variable power argon-ion laser is used for this technique. Typically, the power
required is 750 milliwatts.[4] The laser light sheet can be manually scanned across the
model and can be oriented in either the horizontal or vertical direction.

EXAMPLE TESTS

Trapped Vortex Investigation
      One test conducted in the Flow Visualization Facility provided a quick
illustration of a mechanism's cause for base drag reduction observed in wind tunnel
testing and in flight. If a disk is mounted some distance behind an ogive cylinder, its
base pressure can be reduced. A flow visualization investigation revealed the cause of
this reduction. Figure 7 shows an annular vortex trapped between the cylinder and the
trailing disk. The flow visualization technique used in this investigation was laser light
sheet with aluminium particles acting as a flow tracer. This investigation provides an
example of how flow visualization can be used to quickly understand complex flow
fields.

Fig. 7.   Trapped vortex investigation study.

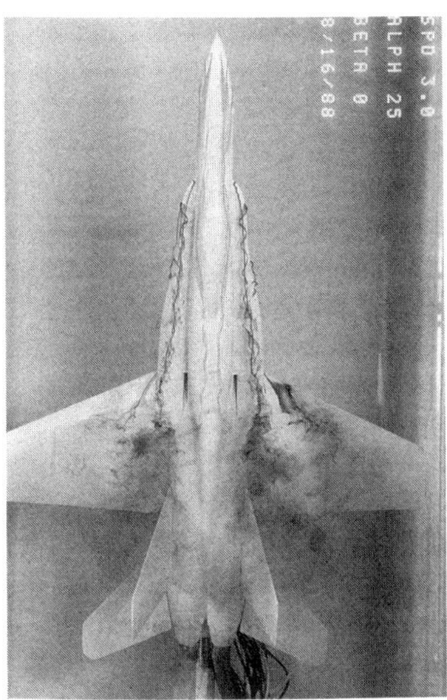

Fig. 8.   F-18 1/32-Scale Model

The F-18 High Angle of Attack Research Program

NASA has been conducting flight tests on an F-18 aircraft to study its flow field at high angles of attack. To support this and other similar efforts, several water tunnel flow visualization tests have been performed on F-18 models built from a variety of commercially available kits. Results from flight, water tunnel, and wind tunnel tests along with computational fluid dynamics results will be correlated. Figure 8 is a photograph of a 1/32-scale model in the water tunnel. The flow visualization technique used here is dye injection.

At an angle of attack of 25°, a pair of leading-edge extension vortices are visible and burst in front of the horizontal tail. Figure 9 shows a comparison of burst-point location as a function of angle of attack. Results were obtained in flight and during several water tunnel and wind tunnel tests.

There is a good correlation between flight results and those from wind tunnels and the 1/32-scale water tunnel test. A water tunnel test using a 1/48-scale model shows poor comparisons with the rest of the tests. It is believed that the 1/48-scale model does not have the correct geometry in the nose area and wing leading edges. The results of this investigation were encouraging in that they compared well with in-flight measurements.

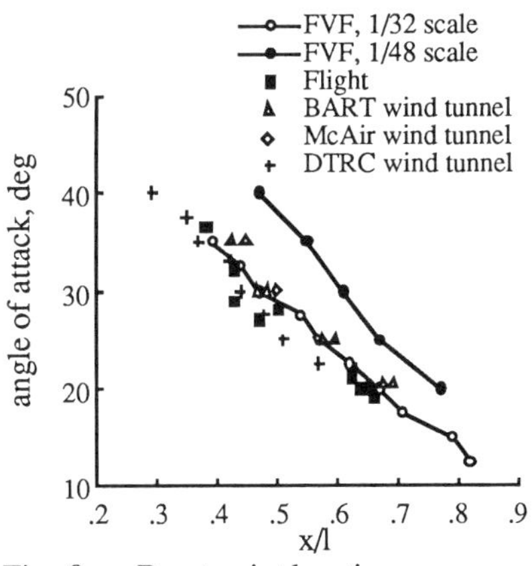

Fig. 9.   Burst-point location comparisons.

McAir 279 Study

The McAir 279 model is a supersonic V/STOL fighter design. This flow visualization study was a parametric design study. The model was fabricated from scratch and had many removable and interchangeable components such as canards, horizontal tails and wing. The flow field was documented for each configuration of interest in order to see which was most favorable for the design goals. Figure 10 shows the model with canards, wing in the aft position and no horizontal tail. The flow visualization technique used is dye injection. This type study can be very useful in the preliminary or conceptual design phase of a vehicle.

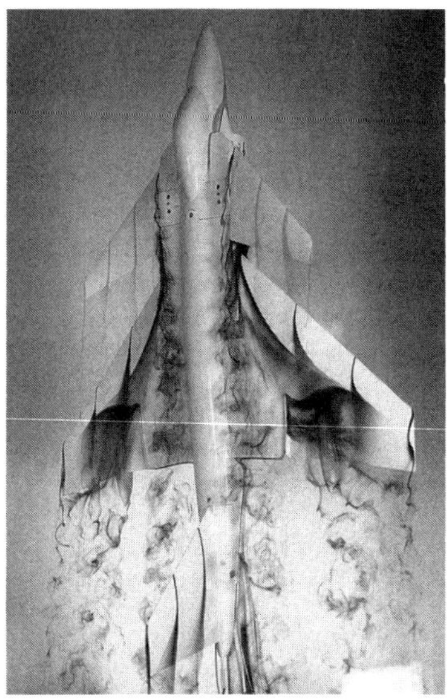

Fig. 10.   McAir 279 design study.

SUMMARY

Through the description of one particular facility, the NASA Ames-Dryden Flow Visualization Facility, an illustration some of the uses of water tunnels is given. Water tunnels can provide a quick and inexpensive means of flow visualization. Although flow visualization does not provide all the answers, it can aid in the understanding of complex fluid mechanics phenomena. Most water tunnel facilities achieve low Reynolds numbers. Thus water tunnels are most useful when the flow field is not Reynolds number dependent such as vortex flow fields. Typically, the results obtained are qualitative, but by using such techniques as laser anemometry, quantitative results can be obtained.

REFERENCES

1. G.E. Erickson, D.J. Peake, J. DelFrate, A.M. Skow, G.N. Malcom, Water Tunnel Facilities in Retrospect and Prospect — An Illuminating Tool for Design, AGARD CP-413, pp. 1-1 to 1-28, June 1987.

2. G.E. Erickson, Vortex Flow Correlation, AFWAL-TR-80-3143, January 1980.

3. F.K. Owen and D.J. Peake, Vortex Breakdown and Control Experiments in the Ames-Dryden Water Tunnel, AGARD CP-413, pp. 2-1 to 2-10, June 1987.

4. C. Beckner and R.E. Curry, Water Tunnel Flow Visualization Using a Laser, NASA TM-867543, October 1985.

THE SIX COMPONENT MAGNETIC SUSPENSION SYSTEM
FOR WIND TUNNEL TESTING

M.J. Goodyer
University of Southampton, Southampton SO9 5NH, U.K.

ABSTRACT

Some wind tunnel testing is carried out with the models
levitated in magnetic fields. The aim of this paper is to review
some of the techniques which have been developed which might also
apply to the testing of models in a liquid helium tunnel. The
accuracy of the data derived from a wind tunnel test on a
magnetically levitated model is potentially higher than when
mechanically supported because, in the absence of the supports,
their aerodynamic interference is avoided and the shape of the
model is not compromised by the support system. Further, the user
may select a wide range of model attitudes more readily than is
possible with the mechanical supports.

The paper describes hardware and the ways in which the
complete set of forces and moments are exerted on a magnetized
model by electromagnets surrounding the test section. These are
used to control the position and attitude of the model and also act
as a source of information on the model's aerodynamic and inertia
forces. Elements of a control system are outlined.

The development of the equipment featured in this article was
sponsored by NASA through Grant NSG 7523, the British Science and
Engineering Research Council, and the University of Southampton.

FORCES AND MOMENTS

Figure 1 shows a wind tunnel model and a set of 10 controllable electromagnets surrounding the test section and is used to illustrate the way control forces and moments are generated. The model contains a magnet running the length of its fuselage. This magnet may be permanent, or of soft iron magnetised by fields from the electro-magnets, or a superconducting solenoid. Lift force and the weight of the model are resisted by magnets 1, 3, 5 and 7, usually by attracting from above and repelling from below using equal levels of current in all four magnets. Pitching moment is produced by the same electro-magnet set but with unequal currents. For example if the attraction of magnet 3 increases and that of 7 decreases by the same amount, a nose-up pitching moment is created with no change in lift force. A side force and yawing moment may be produced in the same manner using the lateral set 2, 4, 6 and 8. The drag force in the direction of the wind suggested by the arrow is resisted by solenoid 9 attracting the model and 10 repelling.

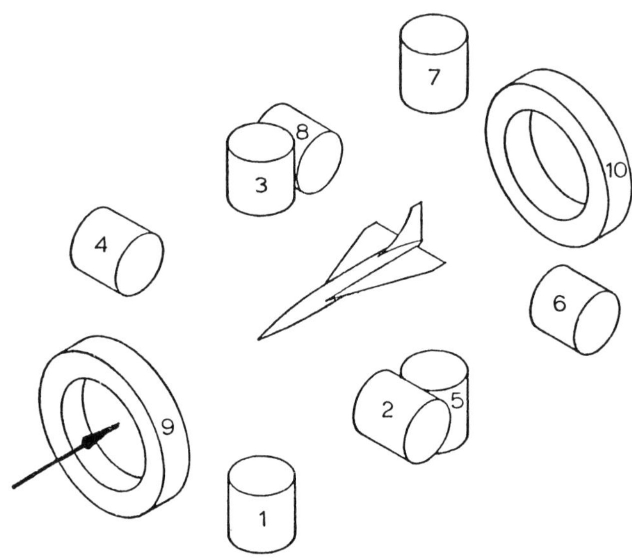

Fig.1. Schematic of the 6-component magnetic suspension and balance system at the University of Southampton.

The array of electro-magnets may be repositioned to some extent to suit the needs of the test, as will be explained in connection with our recent quest to reach 90 degrees angle of attack. This ability to relatively easily modify the system, that is the geometry of the electro-magnets and the model and the

software also to suit the needs of the moment, has been exploited
throughout the development years and has become a feature which may
be listed among its advantages.

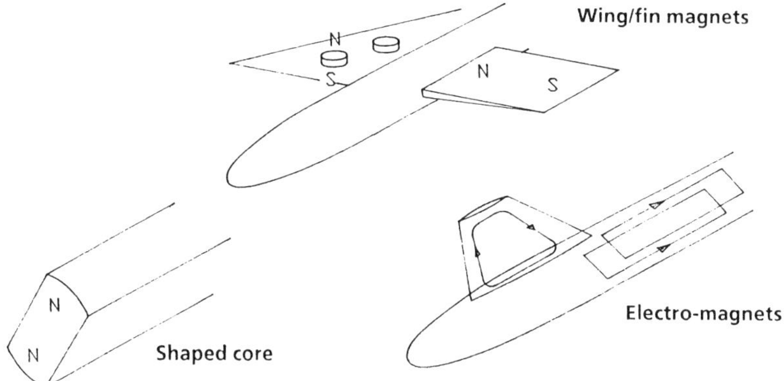

Fig.2. Model elements producing roll control torque.

The methods for producing a roll control moment are less
immediately apparent.   Figure 2 illustrates just some of the
devices which may be built into the designs of models to produce
the moment by interaction with field components specially produced
for this purpose alone by the electro-magnet set just seen.   The
magnet-based devices are the best developed and are now discussed
in turn.

Fig.3. Levitated rectangular-sectioned soft iron bar.

Figure 3 is a picture of a soft-iron shaped-core in
levitation. The core is rectangular in cross-section. Only
electro-magnets 3 and 7 are energised, magnetising the core and
carrying the weight by attraction. The major axis of the core's
cross-section aligns with the local field which is essentially in
the vertical plane through the axis. This confers a roll stiffness
which may be used as a form of passive roll control. The user
cannot adjust the roll attitude with such a simple concept.

The group of photographs in Figure 4 is of a very similar
rectangular sectioned core but this time filled-out to a circular
form by a non-magnetic filler to allow a smooth rotation of the
model in roll. In the left picture it is suspended in the same way
as that just seen: the major axis of the cross-section, which

coincides with the black/white boundary, again is aligned
vertically.

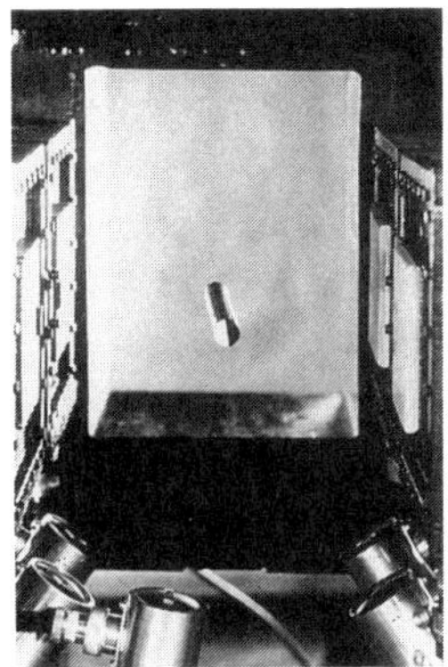 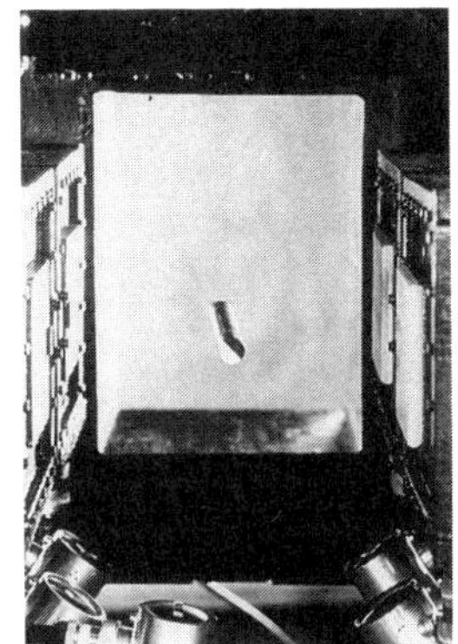 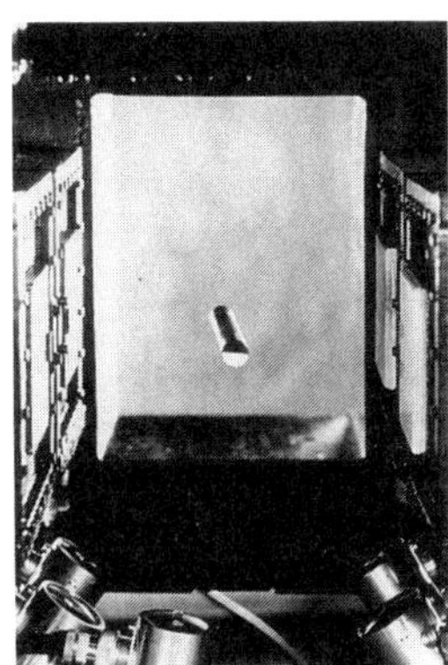

Fig.4. A bar magnet being induced to roll by reorientation of the
magnetic field.

    In the other photographs the model is still supported from
above but with the difference that lateral electromagnets 2, 4, 6
and 8 have been brought into play.   They are energised to attract
the model equally from each side with a field increasing in
strength as we move to the right.   The model is seen to roll,
eventually in the right picture aligning its major axis with the
dominant field from the lateral electro-magnets.   This introduces a
means for actively influencing the roll angle of a model inside a
90 degree range.   The practical embodiment of this idea, used
fairly extensively in wind tunnel testing where closed-loop roll
control was required, took the form of the 45 degree rolled major
axis as normal, the centre picture.   Relaxation of the lateral
electromagnets induces a rolling torque of one sense (clockwise in
this case) by allowing the core to tend toward upright, and the
converse.

Fig.5. A model controlled in six degrees of freedom.

Figure 5 shows a wing-fin-body model. The wings are of cropped delta planform and are in the vertical plane. The electro-magnet array is designed for a high speed tunnel having a rectangular test section which is not present. Rolling motions are controlled closed-loop by the production of a magnetic roll torque using a permanent magnet shaped core in the body of the model, aligned at 45° to the vertical.

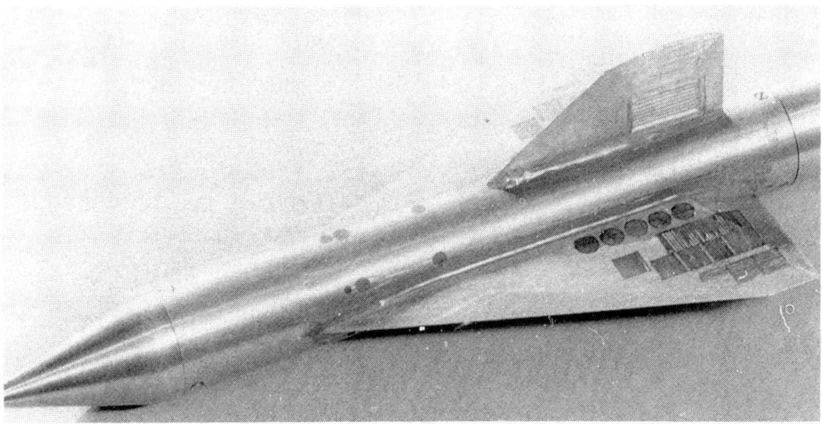

Fig.6. A model tuned to oscillate readily in pitch, used for the
       measurement of pitch rate aerodynamic derivatives.

        Figure 6 shows a delta-winged model designed for closed-loop
control in six degrees of freedom. This also contains a permanent
magnet shaped core running along the fuselage. The core provides
all force and moment components required for control but the moment
capacity in roll is augmented in this case by magnets buried in the
wings, magnetised through the thickness. The position and attitude
of the model are monitored with six light beams. One passes
through a transparent region of the fin to provide roll attitude,
the other five cross the fuselage in appropriate regions to provide
the remaining components of motion. The model has a two-part
construction tuned to resonate in pitch at about 20Hz for the
measurement of aerodynamic pitch rate derivatives.

Fig.7. A model featuring a spanwise magnetised fin to introduce a
roll stiffness.

The field from suspension magnets 3 and 7, which when used
alone will magnetise adequately a soft iron core of a model (when
it is aligned approximately with the axis of the wind tunnel) as
has been seen, may also be used to magnetise an aerodynamic surface
such as a fin when constructed from soft magnetic material. The
fin becomes magnetised in the spanwise direction, aligns itself
with the field and introduces a roll stiffness. The model seen
flying in Figure 7 has four fins, the vertical pair made from mild
steel and the lateral pair from brass. The upper fin dominates as
it was the more strongly magnetised and closer to the active
electro-magnets. The fins may be used for closed loop roll control
in the 45 degree roll attitude, and have been calibrated.

Fig.8. The wings on this calibration model contain permanent magnet
materials for roll control.

The choice of hardware providing a magnetic rolling moment
depends on model geometry. Models with wings and fins tend to need
a high moment capacity and it is natural to use the aerodynamic
surfaces for this purpose. Wings may be magnetised in the spanwise
direction.   This photograph of Figure 8 shows a rolling moment
calibration model (which is not intended for aerodynamic tests)
magnetically suspended and under closed-loop roll control.   With
the electro-magnet configuration of Figure 1 the maximum moment is
available with the wings positioned at 45° to the vertical as seen
here.

ELECTROMAGNET MODIFICATIONS FOR HIGH ANGLES OF ATTACK

It has been mentioned already that the system is changed
from time-to-time to suit the experiment.   A recent example of
major modification of our existing equipment to allow suspension
and force measurement at high angles of attack is partly
illustrated in Figure 9.

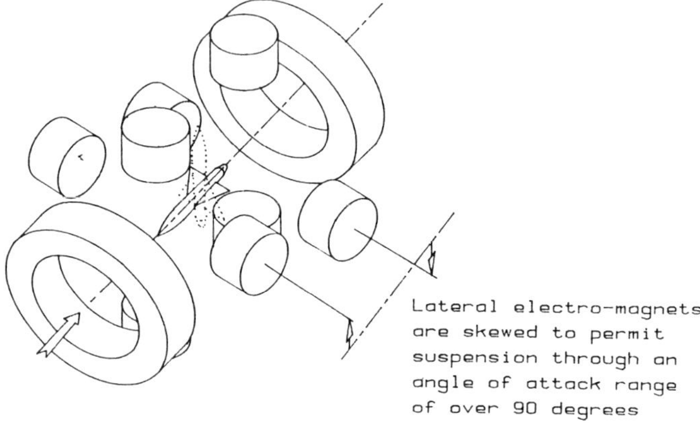

Fig.9. Electromagnet array for high angles of attack.

The four lateral electro-magnets have been skewed to provide sideforce at the high angles. The model shown in full outline is at zero and that in dotted outline at 90 degrees. Other necessary modifications were to the optical position-sensing system to monitor the large angular range, and to the digital controller in order to recognise changes of function of electromagnets with model attitude.

SUPERCONDUCTING MODEL

Calculations had indicated that the magnetic moment of a model containing a superconducting solenoid could be greater than that of a good magnet, even allowing for insulation and helium. The model, shown in section on Figure 10, was built (in our Institute of Cryogenics) and tested in levitation in the electromagnet array of Figure 1. The superconducting coil has 25,000 turns and is rated at 15 amperes. Following the cooling and filling with liquid helium the current was introduced from an external supply, with the superconducting switch open circuit. The model was flown with the switch closed and a constant current flowing in the coil. Gaseous helium escaped slowly from the vent.

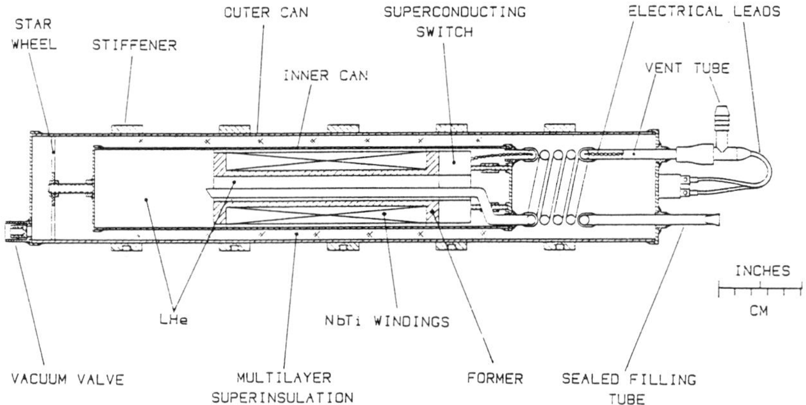

Fig.10. Cross section through superconducting solenoid model.

Such a coil in a model suspended in a liquid helium tunnel could be a particularly attractive proposition. This arises because, with the need for insulation removed, the size of the superconducting coil may be increased relative to the model, increasing its magnetic moment. As a direct result the suspending electromagnets and their power supplies can be appreciably smaller and less expensive.

Fig.11. Launching the superconducting solenoid model.

This model is oversized for the wind tunnel having been designed to prove the concept and for calibration. It is probably the largest to have been suspended in wind tunnel equipment. It is shown on Figure 11 being launched after charging with helium and current and with all umbilicals disconnected. The superconducting lifetime is 30 minutes in this state. There were no operational problems.

FORCE AND MOMENT CALIBRATION

Fig.12. Lift force and pitching moment calibration.

Figure 12 shows a lift force and pitching moment calibration underway with the superconducting solenoid model. Brass weights are hanging just clear of the floor near each end of the model. A pitching moment is applied by using unequal weights. The model is under control in five degrees of freedom with the rolling motion uncontrolled. The attitude in roll is fixed by an offset centre of gravity. An example of a calibration curve derived in this way is shown on Figure 13.

144   M.J. Goodyer

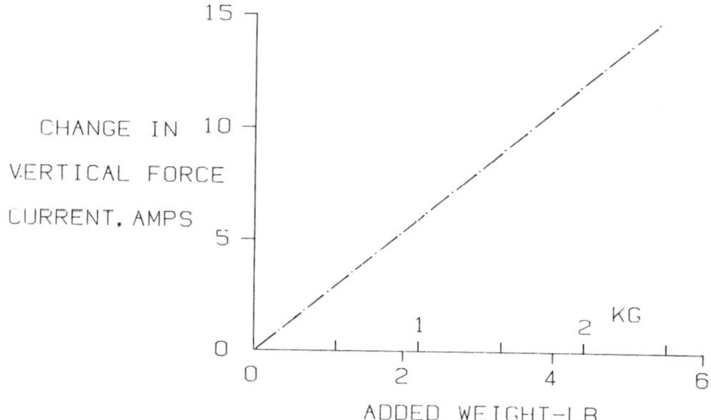

Fig.13. Lift force calibration curve.

The model is that in the preceeding picture and the plot shows the increment in current flowing in each of the electromagnets 1, 3, 5 and 7 of Figure 1 as downforce is applied. The plot is very linear and, although this data is for the superconducting model, is typical in this respect also of permanent magnet models.

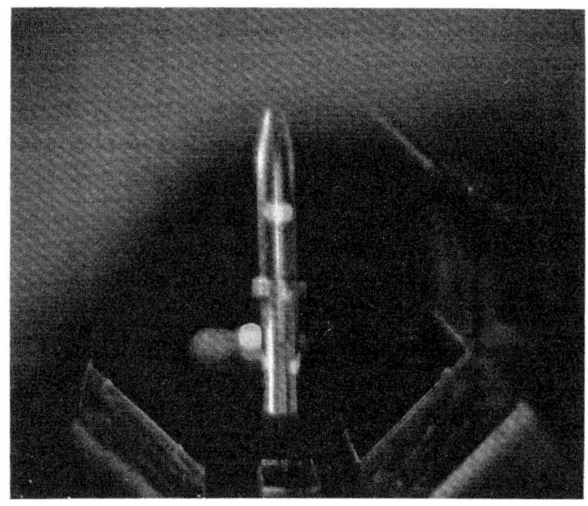

Fig.14. A drag force calibration underway on a model levitated at 90° angle of attack.

Figure 14 shows a drag force calibration underway of an axi-symmetric model suspended at 90° angle of attack. The model contains a neodymium-iron-boron cylindrical permanent magnet core surrounded by an aluminium shell. The drag force is applied through an horizontal thread attached to a ring at the centre of the model. In this attitude the drag force is resisted by electromagnets 1, 3, 5 and 7 of Figure 1 and the lateral position of the model is controlled by electromagnets 2, 4, 6 and 8 but skewed as shown on Figure 9.

CLOSED LOOP CONTROL

The position of a model is unstable, or stable but with low stiffness and damping, depending on the motion being considered. For these reasons normally a closed loop control system is used to stabilise and to introduce acceptable stiffness and damping. The essentials of a control loop are contained in this schematic Figure 15, which illustrates the control of one motion. The elements are repeated as many times as the number of degrees of freedom of the model it is proposed to control.

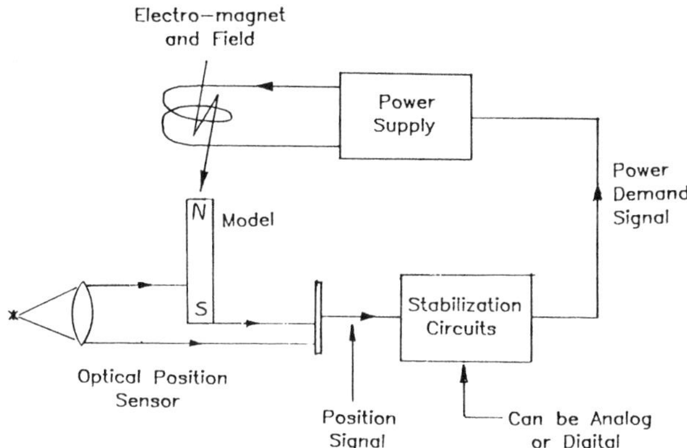

Fig.15. Schematic of control loop for one model motion.

To stabilize the model a position signal must be available. Optical detectors are usual, feeding the stabilization element. The analogue forms of this element include phase advance networks to compensate for a lag in the electromagnets. Proportional and integral components also appear in the power demand signal. Digital versions of the stabilization element are now usual and to date they merely reproduce the analogue equivalents.

Fig.16. Photograph of laser light beams, used for model sensing, crossing the test section.

Light beams of an optical sensing system are shown on Figure 16.   Colimated light sheets from lasers are directed through windows and across the test section.   The model partially blocks each beam when levitated.   The portions of beam which pass the model are directed to linear diode arrays each having 1024 elements.   The position of the model is obtained by counting the illuminated diodes.   In this photograph there are five beams to control five motions of the model with roll excluded.   One beam appears in this view to be vertical and monitors the tail.   The other four cross the test section at 45°.   The outputs of the arrays are manipulated by the control computer on-line to give the three bodily motions and two rotations.   There is no model present: in these circumstances the signals automatically demand zero current from the power supplies.   On launching a model by hand, the presence of the model is detected and appropriate current flows to the electromagnets.

CRYOGENIC WIND TUNNEL WITH MAGNETIC SUSPENSION

An existing cryogenic nitrogen wind tunnel was modified for the magnetic suspension system by designing a new test-section leg. The Reynolds number available at cryogenic nitrogen temperatures is about 12 times that available at room temperature at the same air speed. The essential change to the tunnel was to provide adequate windows in the test section walls for the model position detection optics, in terms of number, position, size and freedom from condensation. An outline of the tunnel circuit and the magnetic suspension system is shown on Figure 17.

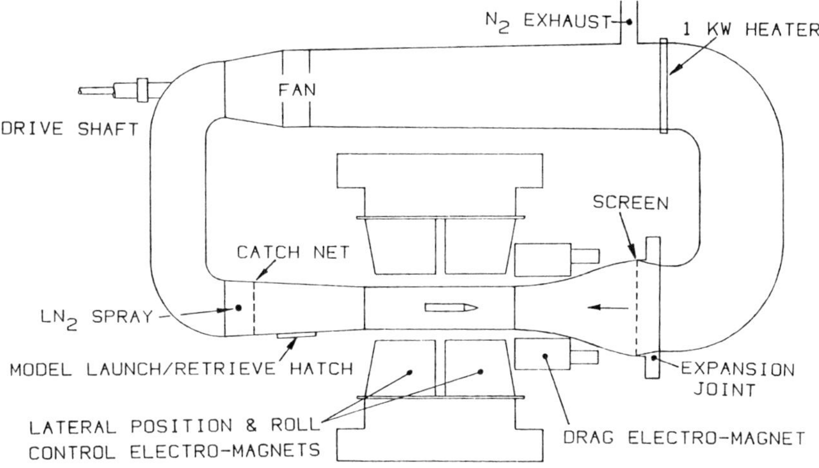

Fig.17. Schematic of 0.1m cryogenic nitrogen wind tunnel with magnetic suspension system.

The project was entirely successful, demonstrating once again the basic versatility of the magnetic suspension system for wind tunnel testing.

Fig.18. The cryogenic tunnel in operation at about 100K.   A model
is flying (not visible here), its drag force being measured.

This photograph, Figure 18, was taken during a tunnel run at
low temperature.   The circular fan section is on the left,
uninsulated and coated on the outside with frost.   The flow here is
away from the viewer, then turns right and returns along the test
section which is out of sight behind the electromagnet set.   The
dark curved pipe is the insulated LN_2 supply to a spray.   Models
were launched by hand through a hatch in the diffuser sidewall just
upstream of the spray.   This is a low speed atmospheric pressure
tunnel operating between about 380K and 80K and reaching a maximum
Mach number of 0.25 when cold.   The test section is octagonal,
approximately 5 inches across flats.

EXAMPLES OF AERODYNAMIC MEASUREMENTS

From the wide range of measurements that have been made on a
variety of models in this facility, just three have been selected
for discussion.   The drag measurements on Figure 19 were made on a
simple aluminium body of revolution containing an Alnico permanent
magnet core.

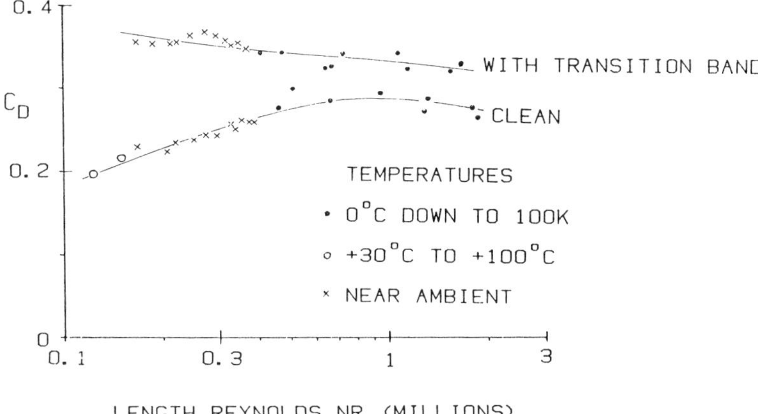

Fig.19. One example of drag coefficient data from magnetic suspension system measurements.

The data is presented in the form of drag coefficient C_D varying as a function of Reynolds number and the surface finish of the model. The spread of Reynolds number was obtained by using cryogenic and high temperatures in the tunnel shown on Figure 18. The increase in drag coefficient at Reynolds numbers up to about 10^6 for the clean model is probably associated with transition moving forward. The transition band, applied to induce a turbulent boundary layer, comprised a 0.2" wide band of size 60 grit centred $4\frac{1}{4}$" upstream of the base. The transition band itself has drag which has prevented the two curves from merging at high Reynolds number.

Drag measurements presented in coefficient form for a clean axi-symmetric model aligned with the flow in atmospheric tunnels are shown on Figure 20. Mach and Reynolds numbers vary with speed. The model and magnetic suspension equipment were common to the tests but the group of points up to M = 0.1 were obtained in a low speed fan-driven tunnel and the rest in an induced flow high speed tunnel, both working at atmospheric stagnation conditions.

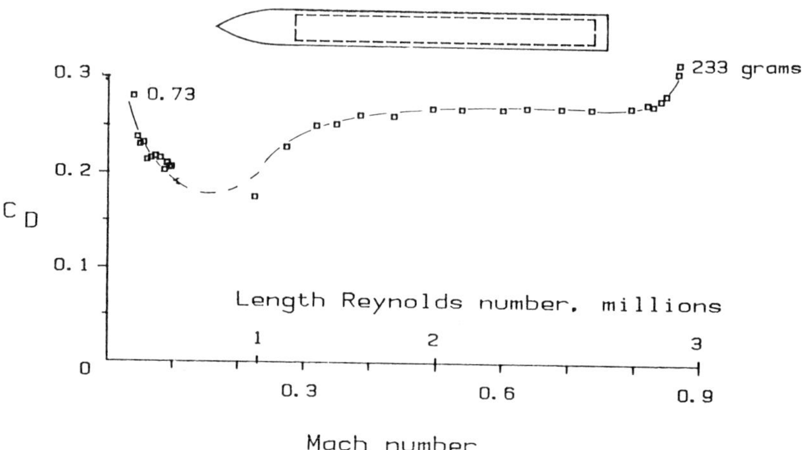

Fig.20. Drag measurements on one model magnetically suspended in two different wind tunnels.

The drag forces experienced at the two extremes of speed are indicated, showing a force range of over 300:1 which was covered using two methods of measurement. The low forces were determined by not controlling the axial motions of the model, leaving it free to deflect axially. For these measurements therefore two motions were uncontrolled, roll and axial. Each had naturally occurring stiffnesses which rendered the motions stable although lightly damped. The axial deflections were monitored optically wind-on and during separate force calibrations, giving good force resolution. At the higher speeds the axial position of the model was under closed-loop control, the drag forces being determined from electro-magnet current. Separate measurements indicated that the boundary layer over the model was laminar at all speeds up to Mach 0.1.

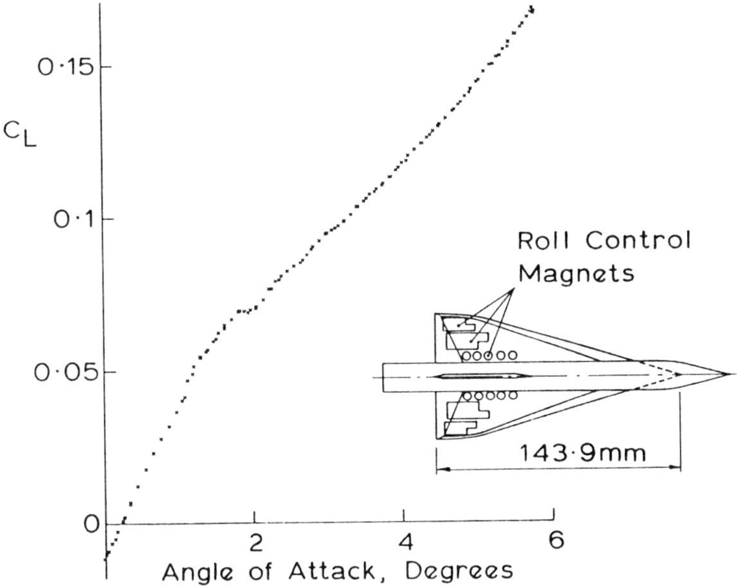

Fig.21. The variation of lift coefficient as a model is ramped
slowly through a small pitch range.

An example of lift measurement is shown on Figure 21. The
model was suspended under six-component control with its wings
vertical in the manner of Figure 5. The roll control moments were
provided in this case only by wing magnets. To obtain the above
data the model was swept continuously through the range of angle of
attack at a rate of 0.5 degrees per second, with simultaneous
recording of the currents in the electro-magnets. It is seen that
fine detail is available in the variation of the aerodynamic
behaviour of the model.

CLOSING COMMENTS

In the aerospace field the product of the research and
development effort by various organizations aimed at improving the
technology of magnetic suspension systems for application to wind
tunnel testing, largely led by NASA Langley Research Center, has
brought suspension and measurement techniques to the point where
models such as that shown on Figure 22 may be reliably tested.

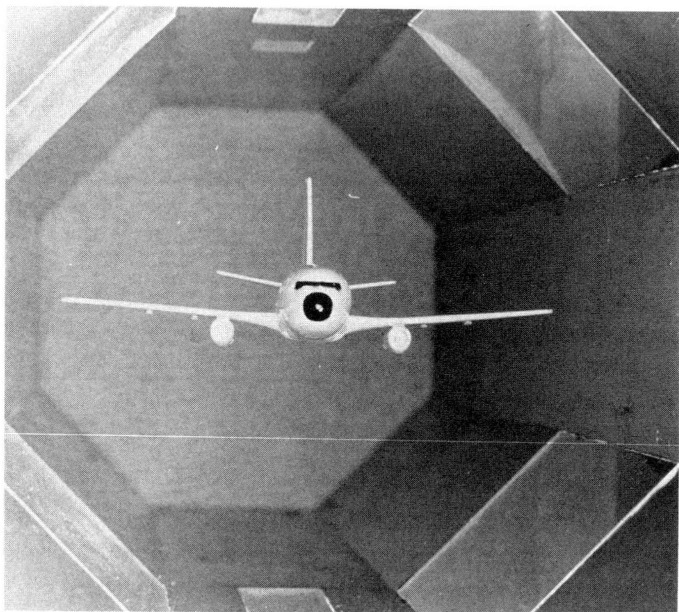

Fig.22. A civil transport model magnetically suspended in a wind
tunnel.

The normal range of measurements is made such as the full
set of forces and moments, (and pressure, as will be discussed on
other papers) in an environment free from the aerodynamic effects
of support interference and compromise in model shape, and free
from constraints on the choice of model motion imposed by
mechanical supports. Superconductivity may be exploited in the
model or levitation coils. The technology has been demonstrated in
the cryogenic nitrogen wind tunnel environment and is used on a
regular basis in various conventional tunnels covering the speed
range from the low subsonic to hypersonic.

MAGNETIC SUSPENSION - TODAY'S MARVEL, TOMORROW'S TOOL

Pierce L. Lawing

INTRODUCTION

There have been more than a dozen Magnetic Suspension Systems (MSS) built and used during the past 30 years (references 1, 2 and 3). There are now six systems in use around the world. All are small, and generate inadequate levels of Reynolds number for simulation purposes. Because of advances in supporting technologies, and the continued advocacy and development of MSS by Langley Research Center, it is now possible to build MSSs for large tunnels. Advances in superconductivity, large magnet construction, computers, automatic control techniques, and innovation in MSS design, make the use of MSS with large tunnels both possible and very attractive. References 4, 5, and 6 deal with the systems aspect of integrating MSS into large tunnels. Reference 7 updates the status of MSS research.

Several speakers at the AIAA 14th Aerodynamic Testing Conference strongly recommended using MSSs to upgrade existing wind tunnels. Several papers seeking to define the best wind tunnel for various speeds called for inclusion of a MSS (see references 8, 9, 10).

It is now possible to generate very high Reynolds numbers in small facilities by the use of superfluid Helium. This technique will be presented by others at this workshop. This paper describes how MSS solves many of the problems caused by support interference in wind tunnels. It proposes significant increases in tunnel productivity through the use of MSS. It also forecasts new test techniques to give us innovative and unique research capabilities.

IMPROVED MEASUREMENT OF STATIC AERODYNAMICS

Eliminate Errors Due to Flow Distortion.

Researchers were aware of support interference in the earliest wind tunnels. Partial corrections for some of these interference problems have been developed. However, there are still secondary errors caused by distortion of the flow which are not easily correctable.

The need to measure base pressure to correct the drag data is well recognized. Even in the most favorable cases, there are interference errors in base pressures measured on a sting mounted model. Also, when the model is at a high attitude the interference error becomes larger. This is because the exposed support causes increased flow distortion at the model support juncture.

Support errors increase when there are large areas of separation with a coherent structure. Two examples of such flows are transverse cylinder flow and the flow over swept delta wings at high lift. The coherent structure in transverse cylinder flow is the oscillating Karman vortex street. In the flow over a delta wing it is the stationary vortex over the wing. Intrusion of the model support into either of these flow fields can cause gross errors in the static aerodynamic coefficients.

Figure 1 shows an example of a cylinder-like flow field which occurs on a missile at high angle of attack. The measured coefficients differ by as much as 40 percent depending on support location.

Figure 2 (from reference 12) shows a comparison of three schemes of supporting a sharp-edged delta wing model. The rolling moment coefficients as a function of sideslip angle differ by as much as 400 percent depending on the support scheme.

Another source of error due to flow distortion may occur at transonic speeds. Figure 3 (from reference 13) shows an example of this problem. Since there is no support free data with which to compare, the problem is shown as the difference in axial force coefficient for several support arrangements. As the Mach number increases through transonic speeds, the errors are not only large, but they reverse direction. Reference 14 gives additional information on support interference problems that can be solved by using MSS.

Eliminate Error Due to Geometry Distortion.

Figure 4 illustrates distortion to the model geometry for sting clearance. This distortion was also needed for dynamic-stability tests of the Shuttle Orbiter-747 combination using a conventional sting support system. One of the concerns of the Shuttle Orbiter-747 combination was the effect of the blunt base of the Orbiter on the performance of the 747 vertical stabilizer. Enlarging the fuselage of the model of the 747 to clear the sting not only submerged part of the vertical tail, it also altered the flow on the horizontal tail.

Remove Restrictions On Static Testing Due to Supports.

Static testing on conventional sting support systems is often restricted in angle of attack. The usual solution is to use a bent or "dog leg" sting to keep the model away from the test section ceiling. Even with bent stings, large angles of pitch or yaw often require the use of an extra, smaller model. A MSS eliminates the mechanical support system. You can freely position the model in the center of the test section with no restrictions due to supports.

A more complex restriction occurs in testing a model in ground effect, either during takeoff, landing, or hover. Figure 5 is a photograph taken during tests of the Shuttle Orbiter in a landing attitude. To obtain the data it was necessary to build an artificial ground plane in a large tunnel and then use a small model. Notice the slot cut through the ground plane for the sting. A later phase of this test required cutting additional slots to obtain data for cross-wind conditions. A MSS requires none of these devices to get the data. Rather than test at a few discrete angles, a continuous sweep of cross-wind angle is available with a MSS.

Improve Balance Accuracy.

The accurate measurement of the forces and moments acting on the model is of paramount importance to the aerodynamicist. Of special concern is the ratio of lift to drag, L/D. The aim of configuration synthesis is to minimize the trimmed drag at the angle of attack for maximum lift to drag ratio. The drag, being the smaller force, is the most difficult to measure accurately. A MSS measures current to determine the drag force, and current can be measured very accurately. Also there are no temperature gradients or gage heating problems. Thus, MSS

has the potential to make very accurate drag measurements, much more accurate than conventional strain-gage balances. Fidelity of models and flow quality will be the limiting factors in MSS measurements. Reference 15 discusses drag measurement in the Langley 13" MSS.

IMPROVED MEASUREMENT OF UNSTEADY AERODYNAMICS

One of the major potential advantages of MSS is its ability to give accurate dynamic stability data, either in a single or multiaxis mode. It is this advantage that originally led to the interest in MSS at NASA Langley in 1957.

Provide Improved Dynamic Stability Data.

As we have seen, in small-amplitude dynamic-stability testing it is often necessary to enlarge the rear of the model to clear the fixed sting. Also, the typical forced oscillation mechanism severely limits the location of the center of oscillation. This has limited the usefulness of test results on configurations such as the Mercury and Gemini Capsules and the Apollo Command Module.

In contrast, with MSS there is no distortion of the model to accommodate the support system. With MSS we can vary the oscillation center at will. This is a very desirable feature for configurations where the center of gravity is unknown or where there is a large shift in center of gravity over the flight envelope.

Provide Multiaxis Capability.

Another advantage of MSS for dynamic-stability testing is the ability to have simultaneous motion of the model, either forced or free, in all six degrees of freedom. For example, this capability can produce a combined pitching and plunging motion to give a pure angle of attack oscillation. This particular motion can be used to separate the effects of Cm_q from the effects of $Cm_{alpha\text{-}dot}$.

Improve Parameter Identification Techniques.

The development in recent years of parameter identification techniques to determine the dynamic characteristics of physical systems fits in well with a MSS. We can use parameter identification techniques to determine the dynamic stability characteristics of models in a MSS. Even when the model appears to be stably suspended, it is always moving a small amount. The currents in the supporting coils are changing as required to maintain the model location. The dynamic characteristics of the model are extracted from the currents. The amount of model movement can be increased by reducing the control system gains. This may improve some testing, for example, a close coupled canard configuration. Data can also be acquired from the position sensors and, by telemetry, from accelerometers inside the model.

INCREASED WIND TUNNEL PRODUCTIVITY

Increased wind tunnel productivity is an area where MSS has an especially strong potential. We increase productivity if:
1. the required data are obtained with less tunnel occupancy hours,
2. fewer models are required,
3. less data reduction is required
4. the data per hour are of higher quality,
5. previously unavailable, but desired, data are produced.

The following examples explain features of a MSS that can increase productivity.

Reduce tunnel occupancy.

Reference 16 describes a dynamic-stability test of the crew escape module of the B-1 bomber in the Langley 8-ft Transonic Pressure Tunnel. Such large tunnels are expensive to operate. Also, their time is valuable since they have more customers than they can serve.

It took five model/sting combinations to get the data. It was necessary to oscillate the model through five ranges of angle of attack in pitch, four ranges of yaw angle, and at four roll positions. The test section opened many times to change the model and the support system.

The test would have been much different if there had been a MSS. It would have taken one less model. There would have been no distortion of model geometry. There would have been no supports to build. The data would have been free of support interference. There would have been no need to open the test section. The testing time would have been a few days rather than several weeks.

Eliminate support interference studies.

Figure 6 (from reference 17) shows one way to evaluate support interference. In this example, the evaluation was necessary to allow extrapolation of the wind tunnel results to flight. Figure 7 shows the size of the corrections determined and their nonlinearity with Mach number. For a typical cruise condition, these corrections are about 10 percent of the total drag. This sizable correction is essential if the data are to be used for configuration synthesis and performance prediction.

In terms of productivity, the use of MSS would have resulted in savings in three areas: First, the large family of expensive support hardware sketched in figure 6 would not have been built. Second, additional tunnel time to test the different support configurations, and the test section entries required to make configuration changes would have been eliminated. Third the extensive data analysis procedure documented in reference 17 would have been avoided.

Improve the use of existing test techniques.

Figure 8 shows a wind tunnel model of the Shuttle Orbiter attached to a ferry ship. It is important to determine the aerodynamic coefficients of each vehicle in terms of the Orbiter attitude, axial location, and separation distance. The conventional way to get this data is to test using several values of each variable. This requires many different attachment points.

Using a MSS would allow the test engineer to fly the Orbiter to different locations relative to the ferry. He would simultaneously measure the forces on the ferry with a mechanical balance and on the Orbiter with the magnetic balance. This would let him determine the sensitivity of the aerodynamic coefficients to location and attitude with one test set up. This procedure would reduce tunnel time, reduce the number of test section entries, and eliminate the time taken for model changes. It would also give a higher quality, previously unavailable answer, thus improving productivity.

One way to study store separation is to support each model with a force balance. This allows simultaneous measurements of the aerodynamic forces. With a complicated drive for the support of the model of the store, it is possible to simulate the positions a store would take during a separation, and then determine the mutual interactions between store and aircraft. References 18 and 19 summarize the practice of this art.

In a tunnel fitted with MSS, the aircraft model would be built of a non-magnetic material, such as stainless steel. Its support and balance would also be non-magnetic. The model of the store would contain a magnetic core and be magnetically supported and positioned. This would allow greatly increased productivity in stores testing. The use of magnetic fields as an aid in stores release testing was discussed at least as early as 1967 (see reference 20).

NEW TEST TECHNIQUES

The previous sections of this paper have dealt with at least quasi-conventional uses of a MSS built using existing technology. This section describes advanced MSSs equipped with systems based on advanced technology.

One of the critical areas in using a MSS is position sensing. A potential improvement over existing position sensing systems is a small inertial guidance system just coming to market (reference 21). It is about the size of a golf ball, small enough to fit in a model for a large tunnel. A high speed computer will use information from the inertial guidance system, together with information from the conventional sensors, for improved control and data gathering. A master control computer will oversee high speed microprocessors and power amplifiers for each magnet.

These new control systems will freely manipulate a model in pitch, roll, and yaw, or combined motions such as spin or tumbling. Also predicted is the ability to relinquish control of any or all degrees of freedom for intervals of time so the model can dynamically respond to the aerodynamic forces. For example, releasing the roll control would allow a fighter model at angle of attack to oscillate in roll in response to the forces that induce wing rock. Reference 22 discusses several potential applications. Some of these are updated here along with new examples.

Fighters

Testing of fighter models has always offered the wind tunnel experimentalist a challenge; a straightforward problem is the high angle of attack required. To maintain force balance measurement accuracy with increasing angle of attack, it may be necessary to change balances

several times since the ratio of beam loads can change by a factor of two or more. Also, it is usually impossible to use a single support, and it may be necessary to use one or more bent stings. Finally, at the higher angles and with the attendant massive separated wake, some of the aerodynamic coefficients become very sensitive to the presence and type of support structure, as discussed in reference 12.

The use of an advanced MSS could eliminate these problems and simultaneously add new capabilities. Once at high angle of attack, a fighter may exhibit the tendency to buffet, wing rock, nose slice, or in some other way depart from a stable condition. Since you can relax or abandon any constraint on model motion imposed by the MSS, you can study these tendencies, or artificially induce them and allow the motions to amplify or decay. In addition, we can determine high angle of attack characteristics instantaneously. This will allow us to measure, for example, the overshoot in lift coefficient that occurs as a result of the pitch rate needed to get a fighter to the high lift condition.

Testing a fighter model at high angle of attack is one example of the new capabilities of an advanced MSS. Others are pilot-in-the-loop, magnetically varied center of gravity location, real time computer generated stability augmentation systems, computer generated artificial damping, out of flight path pointing, simulated carrier landings, and simulated refueling. We leave it to the reader's imagination to come up with more examples of using a MSS to solve a particular problem in fighter testing.

Missile Separation

The technique of studying two-body or stores separation problems with a MSS has been discussed above. The following example provides the philosophy for additional uses of advanced MSS for missile problem solving.

Figure 9 illustrates a test in the 13" MSS. A cavity, which simulates a bomb bay, is supported by the "traversing support arm" shown in the figure. By using drive screws through the support arm, the cavity can be raised, lowered, or pitched. The cavity walls are made of glass to avoid interference with the laser positioning system. The MSS holds the store stationary on the test section centerline while the cavity is raised, thus simulating a store dropping from the cavity. Also, the cavity can be fixed and the store moved out of the cavity in small increments.

The purpose of this test was to study the unsteady loads in the cavity and during separation rather than find the separation trajectory. The first phase of this test demonstrated that the MSS could support and control the missile in the cavity, partially out of the cavity, and in the neighborhood of the cavity. We were able to decrease the control loop gains and let the missile move in response to the unsteady loads.

Figure 10 is a photograph of an ogive-cylinder missile model in suspension within the cavity. In future tests we will sample the forces and moments several hundred times per second. Once we determine the aerodynamic loads, we will oscillate the store at various stations as it emerges from the cavity. We can also vary the frequency of oscillation to determine cavity resonance conditions.

With a special coil arrangement, a cavity in the tunnel floor could function as a magnetically controlled model injection system. This provides simulation of a bomb bay or silo launch. If we had this system in a trisonic tunnel, we can envision a special naval problem of shipboard defense as a potential application.

Since the direction from which a threat might approach is not known, we can assume the missile launcher would be mounted vertically. The most demanding task from such a launcher is interception of a sea level threat. Figure 11 shows the missile first being moved magnetically from a simulated launch tube to simulate this mission. Air blowing from the tube would simulate booster exhaust. This portion of the test could be repeated at different tunnel speeds to simulate different levels of crosswind.

The next phase of this type of launch is a high g turn, requiring the missile to fly at very high angle of attack to the oncoming flow. (Since the typical missile of this type has no wings, it turns by having a component of thrust normal to the flight path, and the angle of attack is negative in the aerodynamic sense.) During this maneuver, the Mach number, pitch rate, and angle of attack are all changing. By using the computer controlled stop point trajectory technique discussed below, we can study these variables one at a time or all together. The final phase is supersonic lock-on to the target. Static stability, dynamic stability, and control force requirements are known at every point in the trajectory.

Computer Controlled Trajectories

Figure 12 illustrates the concepts of "computer controlled stop point trajectories," and "designer in the loop." This example is of a crew escape module being ejected from a military aircraft. Here, the aircraft model is mounted on a non-magnetic but otherwise conventional blade mount. The crew escape module contains a magnetic core and is manipulated by a MSS. The purpose of the test will be to find escape module trajectories that allow the crew to survive.

The test would be conducted by first testing with the crew module in place on the aircraft. This gives the initial aerodynamic separation forces tending to hold or eject the module. Tare forces due to friction are found for this first data point by moving the module very slightly and retaking the data.

A non-aerodynamic initial force and trajectory consistent with an active ejection mechanism, for example a rocket is assumed. This initial trajectory is the one the module would follow in a vacuum.

The next point is obtained by moving the module along this prescribed trajectory just enough to allow flow to circulate between the module and the cavity. The measured forces and moments are used to modify the vacuum trajectory and allow the calculation of the next point along the modified trajectory. At the next point, forces and moments are again calculated, and so forth.

The assumed center of gravity of the module is forced by the MSS to follow the modified trajectory. Likewise, the MSS rotates the module as dictated by the measured aerodynamic moments and calculated moments of inertia.

The above process describes the "computer controlled stop point trajectory." This trajectory could be programmed to run many times during a wind tunnel test. It is then possible to make parametric studies of such effects as the attitude of the aircraft, the c.g. location in the module, variations of the initial trajectory, and settings of the aerodynamic control surfaces of the module.

The g loads that the crew would experience are continuously calculated. From the g loads and the trajectories an experienced designer in the tunnel control room could identify successful and unsuccessful ejections on line. He could then introduce perturbations for the initial conditions for the next trajectory. Continuation of this process would define the boundaries of a survivable escape.

We could impose other ground rules such as the module remaining upright, or not spinning, and certainly not hitting the vertical tail of the aircraft. This information is then used to decide the safe operational envelope of the crew escape system.

This is the essence of the "designer in the loop" concept. This technique will give a complete answer in a few hours. Conventional methods take months for a partial answer.

This hypothetical example illustrates how an advanced MSS might be used to improve both accuracy and productivity in wind tunnel testing.

Helicopters

Figure 13 shows an additional capability offered by MSS in the testing of helicopters. The testing shown in the photograph is made with the rotor turning from an attachment in the test section floor. The helicopter body is being tested in the downwash of the rotor but is detached from the rotor. The body rotates 360 degrees to produce crosswinds from the tunnel flow. This will allow the determination of the static stability of the body in the hover mode and under the combined effects of rotor downwash and crosswind.

A MSS is an ideal replacement for the support used here to turn the body. The MSS gives data free of support interference and greatly increased ease of operation. Also, with the set up shown in figure 13, no particular beam of a mechanical balance can be optimized since the maximum load can come from any direction. The magnetic balance components are automatically optimized for each load condition.

Transports/Bombers

Figure 14 illustrates an advanced subsonic transport typical of the "family concept" of more efficient aircraft as described in reference 23. The use of the twin fuselage scheme presents the test engineer with the immediate problem of finding a place for a conventional strain gage balance. A MSS can ease this problem with a magnetic core in each fuselage, creating a virtual core in the center of the model. This also solves the vexing problem of supporting such a model, both for static and dynamic stability tests.

The advanced transport shown in figure 15 presents a different testing problem. The main lifting surface is so far aft that it is impossible to use a mechanical support without severe support interference problems. A MSS not only gets rid of the support problem, but also offers the opportunity to fly this canard dominated design with varying degrees of positive or negative static margin.

Hypersonics

All the advantages of wind tunnel testing with a MSS are retained at hypersonic speeds. The problem of changing the vehicle geometry to accommodate supports becomes particularly acute for airbreathing vehicles optimized for cruise range. To reduce both drag and heating, it is necessary to have highly swept thin wings and stabilizing surfaces, as shown in figure 16. It is obvious that the aft section of this configuration would be grossly distorted by the conventional balance and sting.

At the present stage in hypersonic research, fundamental studies of wing alone aerodynamics are needed to develop optimization methods. Since the wings are normally thin, there is no place for a mechanical balance. High quality data is easy to obtain with a magnetic balance. For these configurations, the entire wing would be made of a magnetic material in order to be suspended by a MSS.

Figure 17 is a photograph of a "parasol wing" configuration. This is a two-body configuration since the wing is supported on pylons and there is flow between the body and the wing. The performance of the wing is improved in "wave rider" fashion by the compression of the flow of the fuselage forebody, and the drag of the fuselage is reduced by the aft body high pressures induced by the wing compression.

A MSS would be used here in several ways. The MSS would first measure the wing alone performance of the very thin wing and then determine the best location of the wing. A non-magnetic wing would then be blade supported and the fuselage magnetically flown to determine a similar data set. Finally, the entire optimized configuration would be tested magnetically to measure its aerodynamic performance.

Hypersonics has other special problems that are solved by using a MSS. One of the problems of conventional balances which may occur in hypersonic tunnels is high model heat transfer rates. These can induce large temperature gradients in the balance. In the hypersonic environments essential for simulation of turbulent boundary layers, these gradients are large enough to require water cooling of the balance. In a few cases even water cooling cannot reduce balance temperature gradients to acceptable levels. Use of MSS will eliminate the temperature gradient problem.

It is obvious that MSS holds great promise to improve hypersonic testing. In fact, some of this promise has already been realized. The bibliography (reference 2) lists 6 tunnels that have used MSS in hypersonic testing. Mach numbers ranged from 7 to 16. A table of MSS tunnels in reference 3 lists 7 tunnels with Mach numbers from 5 to 16. The tunnel at Oxford University is listed as active.

A NEW TUNNEL CONCEPT

A new tunnel concept is discussed in this paper. It is a combination of three types of facilities: magnetic suspension and balance systems, ballistic ranges, and cryogenic tunnels. We call it a $\underline{M}$agnetic $\underline{F}$light $\underline{T}$ube, or MFT. Figure 18 shows a conceptual drawing of a MFT.

The model is magnetically suspended in a central tube which contains the test gas. The temperature of the test gas can be varied from 100 K to 400 K. Pressure of the test gas can be varied from near vacuum to pressures high enough to simulate flight Reynolds number. A MSS external to the central tube supports the model. The MSS mount is called a cart. The MSS cart is itself magnetically supported in an external tube in a manner similar to magnetically levitated trains. The MSS cart is accelerated down the tube, carrying with it the model in the central tube.

An actuator accelerates the MSS cart. The type of actuator depends on the speed range desired. To decrease drag on the cart, the external tube is evacuated. For low speed testing the cart is accelerated by bleeding atmospheric pressure in behind the cart. For high speed testing the MSS cart is accelerated by linear motors built into the rails shown on the inner wall of the outer tube. The structure that carries the MSS will also carry the necessary computers and data gathering system.

An advantage the MFT shares with the ballistic range is aerodynamically quiet flow. In contrast to the ballistic range, the model in the MFT is fully controlled and force and torque data are continuously available.

Cryogenic operation lowers the speed for a given Mach number. It also increases the Reynolds number, and decreases the model loads for a given Mach and Reynolds number. The ability to control temperature also gives the usual advantages of cryogenic operation: a wide range of Reynolds number and the ability to vary independently aerodynamic parameters. These parameters include dynamic pressure, Mach number, Reynolds number, velocity, and reduced frequency. In addition, at high supersonic Mach numbers, cryogenic operation means lower total temperatures and lower heat transfer rates to the model. Cryogenic total temperatures are not available in a conventional supersonic wind tunnel because the test gas is accelerated rather than the model. Also, the static temperature drop due to the expansion process is lower than the condensation temperature of the test gas for many applications.

The idea of a moving MSS is not new. Reference 24 describes a system used to move a MSS along a set of rails. The purpose was to move a model upstream of observation windows and a survey rake in a hypersonic tunnel. The MSS used x-ray position sensors to see through the aluminum walls of the relatively long test section. Moving the model upstream in the test section allowed detailed study of the wake behind the model. This entire system was later given to the Langley Research Center. Parts of it are still used in the Langley 13" MSS.

The MFT has potential across the speed range. Low speed testing is carried out in a straightforward manner with the usual corrections for wall effects. Transonic testing may be more difficult, perhaps requiring a length of the center tube with a slotted wall and plenum chamber. The MFT is particularly well suited to the supersonic "quiet tunnel" type of test. Since the model is moving and the walls are fixed, shock reflection is less of a problem. Hypersonic testing will provide heat transfer testing capability at high Reynolds numbers. Hypervelocity limits will be set by the engineering difficulties of acceleration and cart speed.

The special capabilities of this new tunnel would provide unique testing capabilities. Take-off and landing could be studied without the need for a moving ground plane. Since there are no supports, data on phenomena such as dynamic ground effect are easily obtained. Unsteady aerodynamic data are available over the entire speed range of the tunnel. Dynamic stability derivatives may be obtained either by oscillatory or parameter identification techniques.

Spin departure maneuvers could be closely simulated. Impulse loads due to flying through a blast wave, rocket plume, or micro-burst would be measured using standard ballistic tube techniques. The high sensitivity of the magnetic balance, the absence of supports, the variable Reynolds number, and very low turbulence have the potential to provide the world's best drag measuring capability.

Supersonic testing is the focus of the following hypothetical MFT. The initial section is for acceleration, the center section for testing and the final section is for deceleration. The center section is one kilometer long. A cart speed of one km/sec and ambient temperature gives Mach 3 test conditions for one second. For Mach 3 and cryogenic temperatures (100 K), the required cart speed is reduced to 0.6 km/sec, and the test time is 1.67 seconds. For a cart speed of one km/sec and a temperature of 100 K, Mach 5 flow will last for one second. Data will be taken at rates of several thousand samples per second. Depending on the model size and assumptions of flow stabilization time around the model, about 10 changes in model attitude can be made during each run.

SUMMATION

In the conservative view, the use of magnetic suspension and balance systems (MSS) in large wind tunnels will eliminate or greatly reduce errors due to support interference effects. MSS will also eliminate restraints due to supports, reduce tunnel occupancy time, reduce the number of models required, and reduce the data analysis required to evaluate support interference.

In the slightly less conservative view, the use of MSS will improve data accuracy, high angle-of-attack test capability, dynamic stability test technique, two-body/stores-release testing, and provide nearly free flight test conditions in many circumstances. The cumulative effect of these improvements due to MSS will be a large increase in wind tunnel productivity.

In a speculative view, the addition of pilot and designer-in-the-loop concepts will improve productivity and also open new possibilities in wind tunnel test technique. Although the research and development needed to bring about large MSSs should pay for itself in wind tunnel productivity improvement alone, the technology spin-off may be even more valuable.

All of the new capabilities envisioned in this paper may not evolve due to engineering difficulties. However, it is clear that new testing capability will be available with the application of MSS. Also, experience teaches that use of a new technology like MSS will produce benefits beyond those predicted in this paper.

REFERENCES

1. **Tournier, Marcel**; and **Laurenceau, P.**: Suspension Magnetique d'une Marquette en Soufflerie. (Magnetic Suspension of a Model in a Wind Tunnel). La Recherche Aeronautique, no. 59, July-Aug. 1957, pp. 21-27.

2. **Tuttle, M. H.**; **Kilgore, R. A.**; and **Boyden, R. P.**: Magnetic Suspension and Balance Systems. A Selected, Annotated Bibliography. NASA TM-84661, July 1983.

3. **Boyden, Richmond P.**: A Review of Magnetic Suspension and Balance Systems. AIAA 15th Aerodynamic Testing Conference, May 18-20, 1988, San Diego, CA. AIAA Paper No. 88-2009.

4. **Bloom, H. L.**; et al: Design Concepts and Cost Studies for Magnetic Suspension and Balance Systems. NASA CR-165917, 1982.

5. **Britcher, C. P.**: Progress Towards Large Wind Tunnel Magnetic Suspension and Balance Systems. AIAA Paper No. 84-0413, 1984.

6. **Boom, R. W.**; **Eyssa, Y. M.**; **McIntosh, G. E.**; and **Abdelsalam, M. K.**: Magnetic Suspension and Balance System Advanced Study. NASA CR-3937, 1985.

7. **Boyden, Richmond P.**; **Britcher, Colin P.**; and **Tcheng, Ping**.: Status of Wind Tunnel Magnetic Suspension Research. SAE 851898, Oct. 1985.

8.   AIAA 14th Aerodynamic Testing Conference, West Palm Beach Florida, March 5-7, 1986: AIAA CP 861.

9.   Barnwell, R. W.; Edwards, C. L. W.; Kilgore, R. A.; and Dress, D. A.: Optimum Transonic Wind Tunnel. AIAA No. 86-0755, March 1986.

10.   Bushnell, D. M.; and Trimpi, R. L.: Supersonic Wind Tunnel Optimization. AIAA No. 86-0773, March 1986.

11.   Dietz, W. E.; and Altstatt, M. C.: Experimental Investigation of Support Interference on an Ogive-Cylinder at High Incidence. J. Spacecraft and Rockets, Vol. 16, Jan-Feb. 1979, pp 67-68.   See also AIAA Paper 78-165, Jan. 1978.

12.   Johnson, Joseph L., Jr.; Grafton, Sue B.; and Yip, Long P.: Exploratory Investigation of Vortex Bursting on the High-Angle-of-Attack Lateral-Directional Stability Characteristics of Highly-Swept Wings. AIAA Paper No. 80-0463, March 1980.

13.   Price, Earl A., Jr.: An Investigation of F-16 Nozzle-Afterbody Forces at Transonic Mach Numbers with Emphasis on Support System Interference. AEDC-TR-79-56, AFAPL-TR-2099, December 1979.

14.   Tuttle, Marie H. and Lawing, Pierce L.: Support Interference of Wind Tunnel Models - A Selective Annotated Bibliography.   Supplement to NASA TM 81909, May 1984.

15.   Dress, David A.: Drag Measurements on a Laminar-Flow Body of Revolution in the 13-Inch Magnetic Suspension and Balance System. NASA TP 2895, April 1989.

16.   Davenport, E. E.; and Kilgore, R. A.: Dynamic-Stability Tests of an Aircraft Escape Module at Mach Numbers From 0.40 to 2.16.   NASA TM-X-72680, April 1975.

17.   MacWilkinson, D. G.; Blackerby, W. T.; and Paterson, J. H.: Correlation of Full-Scale Drag Predictions With Flight Measurements on the C-141A Aircraft - Phase II, Wind Tunnel Test, Analysis, and Prediction Techniques.   Volume I - Drag Predictions, Wind Tunnel Data Analysis and Correlation. NASA CR-2333, 1974.

18.   Arnold, R. J.; and Epstein, C. S.: Store Separation Flight Testing. AGARD-AG-300-Vol.5, April, 1986.

19.   Billingsley, J. P.; Burt, R. H.; and Best, J. T.: Store Separation Testing Techniques at the Arnold Engineering Development Center. AEDC-TR-79-1, March 1979.

20.   Covert, Eugene E.: Wind Tunnel Simulation of Stores Jettison With the Aid of an Artificial Gravity Generated by Magnetic Fields, AIAA Journal of Aircraft, Volume 4, No. 1, 1967.

21.   Killpatrick, Joe: New Developments in the Ring Laser Gryo. Scientific Honeyweller, Vol. 8, No. 1, Fall 1987. ISSN 0196-8440.

22.   Lawing, Pierce L.; Dress, David A.; and Kilgore, Robert A.: Potential Benefits of Magnetic Suspension and Balance Systems. NASA TM-89079, February 1987.

23.   Kayten, Gerald G.; Driver, Cornelius; and Maglieri, J.: The Revolutionary Impact of Evolving Aeronautical Technologies. AIAA-84-2445, Nov., 1984.

24.   Crain, C. D.; Brown, M. D.; and Cortner, A. H.: Design and Initial Calibration of A Magnetic Suspension System For Wind Tunnel Models.   AEDC-TR-65-187, September 1965.

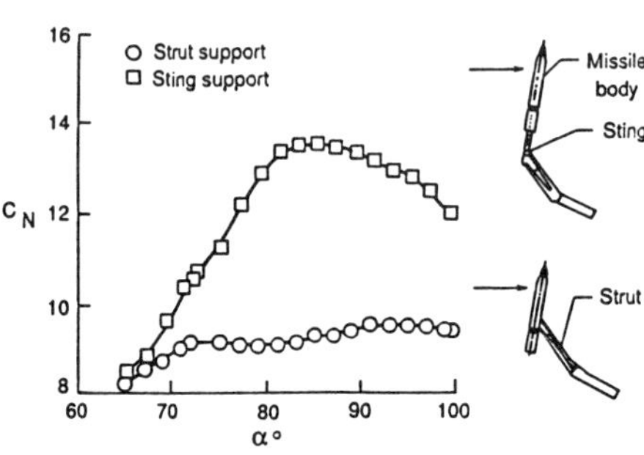

Figure 1.- Normal force coefficients for a missile over the high angle of attack range.

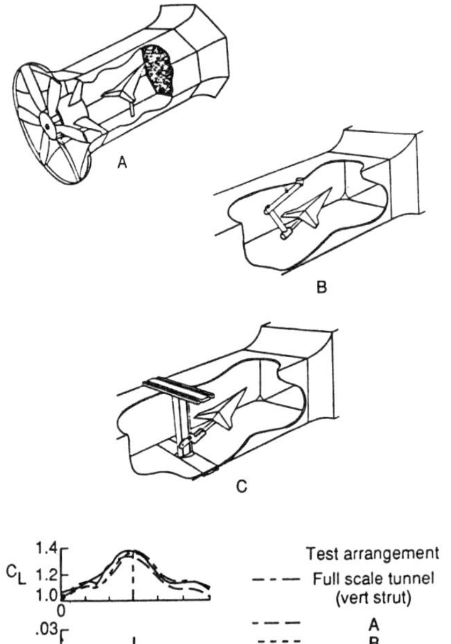

Figure 2.- Support effects on lift and rolling moment as a function of yaw angle.

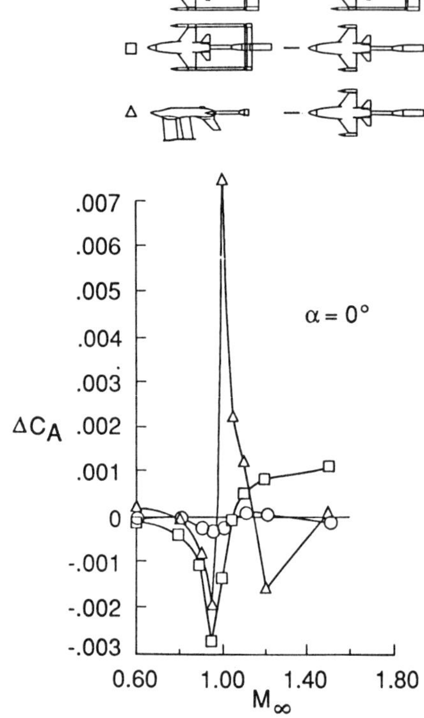

Figure 3.- Axial force increment for three pair of supports as a function of Mach number.

Figure 4.- Comparison of models of orbiter mounted on 747 ferry ship for wind tunnel test and for flight.

Figure 5.- Shuttle landing tests.

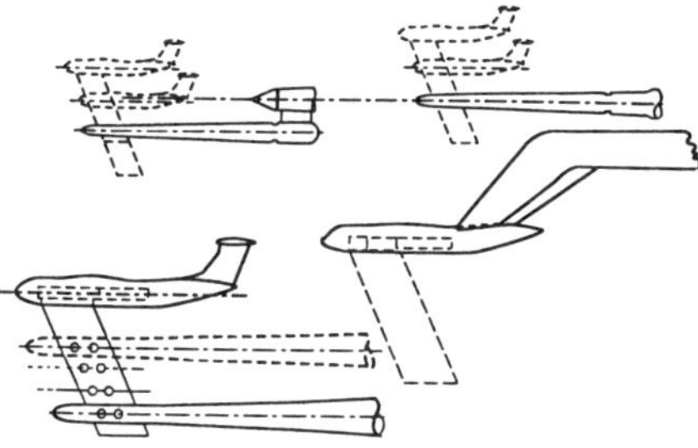

Figure 6.- Support hardware required for support interference tests.

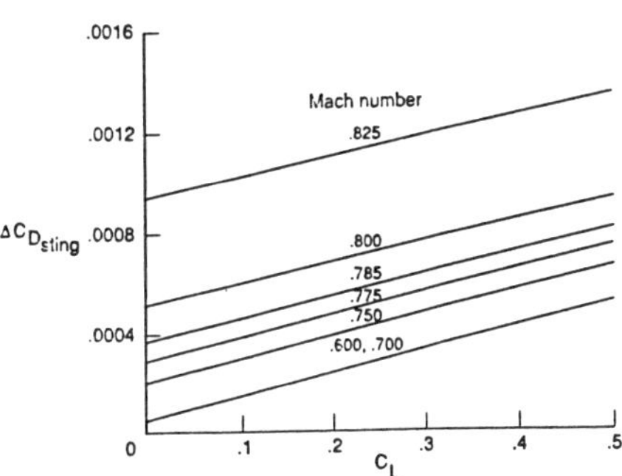

Figure 7.- Drag coefficient corrections for support effects.

Figure 8.- Testing to optimize location of orbiter on ferry vehicle.

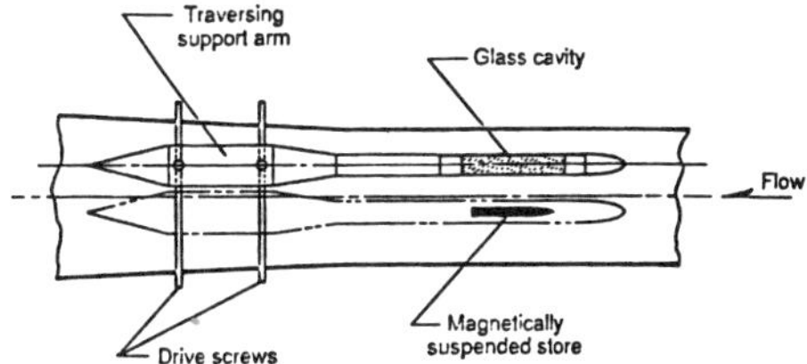

Figure 9.- Magnetic suspension system demonstration of internal stores release.

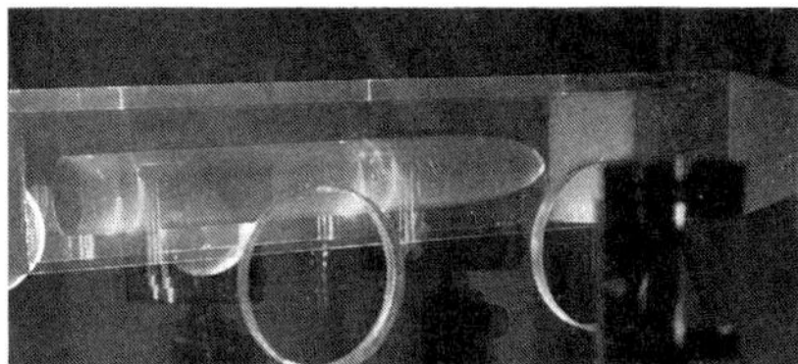

Figure 10.- Missile magnetically suspended in a transparent cavity. (Circumferential streaks are laser beams used for positioning.)

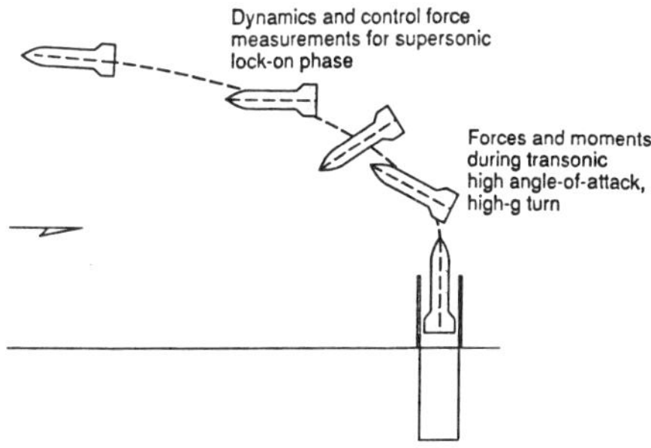

Figure 11.- Conceptual use of a magnetic suspension system to study vertical launches from a ship.

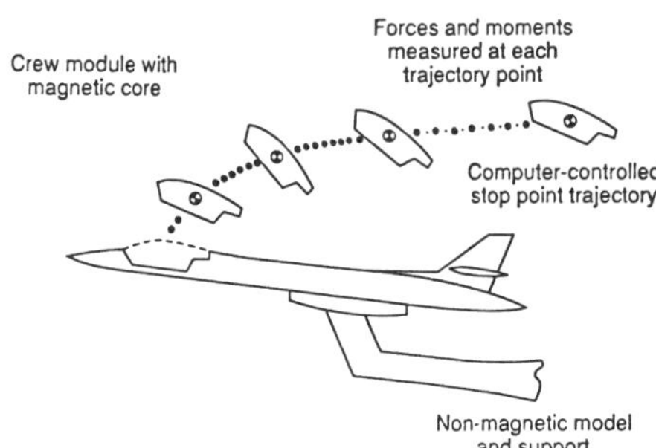

Figure 12.- Conceptual use of a MSS to study escape module parameters.

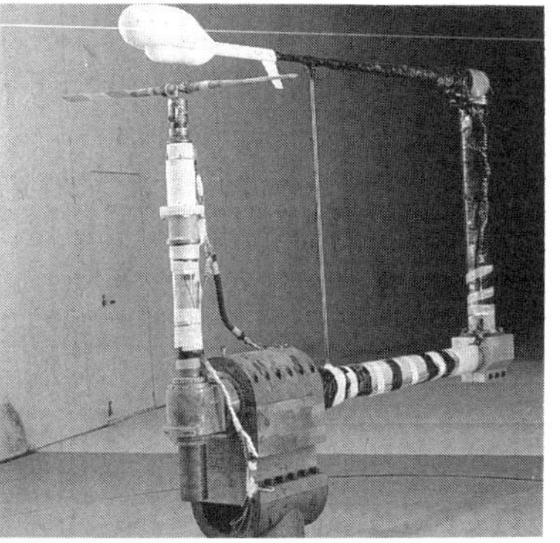

Figure 13.- Helicopter testing for $360°$ horizontal gust response.

Figure 14.- Concept for twin fuselage subsonic transport.

Figure 15.- Advanced subsonic commuter concept.

Figure 16.- Hypersonic bomber concept.

Figure 17.- Parasol-wing hypersonic model.

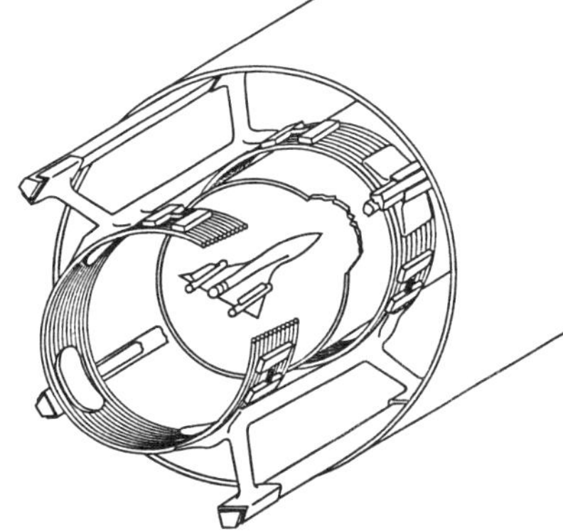

Figure 18.- New facility concept, Magnetic Flight Tube.

Recent Aerodynamic Measurements with Magnetic Suspension Systems

Colin P. Britcher
Department of Mechanical Engineering and Mechanics
Old Dominion University, Norfolk, VA 23529-0247

ABSTRACT

This paper reviews recent aerodynamic tests of a family of slanted-base ogive-cylinders using the NASA Langley 13 inch Magnetic Suspension and Balance System. Results include drag, lift, pitching moment, support interference and base pressure measurements. Mach numbers were in the range 0.04 to 0.2. Drag results are shown to be in satisfactory agreement with previous measurements. Significant support interferences were found at all test conditions. Comparison is made between interference free base pressures, obtained using remote telemetry, and sting cavity pressures. Test results and procedures are briefly discussed in the context of the proposed helium flow facility.

INTRODUCTION

The original interest in slanted-base models was as an analogy to the flow over the sloping rear windows of hatchback and fastback automobiles (Figure 1). For the work reported here, the model is conceptually "inverted" and argued to be an analogy to the flow under the upswept rear fuselage of a transport aircraft. Design constraints are similar in both cases - namely minimum drag and maximum utilization of enclosed volume. The former tends to suggest slender, tapering base geometries while the latter suggest rather blunt geometries. Rather peversely, automobiles frequently show minimum drag with a blunt base. This is due to the formation of intense trailing vortices from the corners of the sloping rear window of fastback designs, leading to high induced drag.

In conventional wind-tunnel testing, the model is mechanically supported in the test section. The aerodynamic disturbance ("interference") caused by the support is a significant problem, with the majority of test results influenced to some extent. Numerous variations of support design are in use, encompassing many types of testing over a wide speed range. The principal type of support used in the transonic, supersonic and hypersonic speed regimes, and the primary focus of this report, is the rear mounted sting. While some information concerning the magnitude of sting interference on drag and base pressures is available for axisymmetric bases, data for non-axisymmetric bases is rather sparse. The slanted-base models thus provide an opportunity to assess sting interference effects over a family of geometries, ranging from axisymmetric to highly non-axisymmetric.

EXPERIMENTAL DETAILS

Slanted-base ogive-cylinder models have been extensively tested by Morel[1] and others[2,3]. The principal interest in this geometry is the sudden change of wake structure, with corresponding large change in drag coefficient, occurring for small changes of base slant angle, around 45° slant. A summary of important previous results is shown in Figure 2. The models used for this research correspond to Morel's geometry and are illustrated in Figure 3; additional details can be found in Reference 4. The models are manufactured from aluminum alloy, with enclosed low-carbon iron cores. Considerable care was taken to ensure geometric fidelity and high quality of surface finish. Interchangeable bases permit a range of base slant

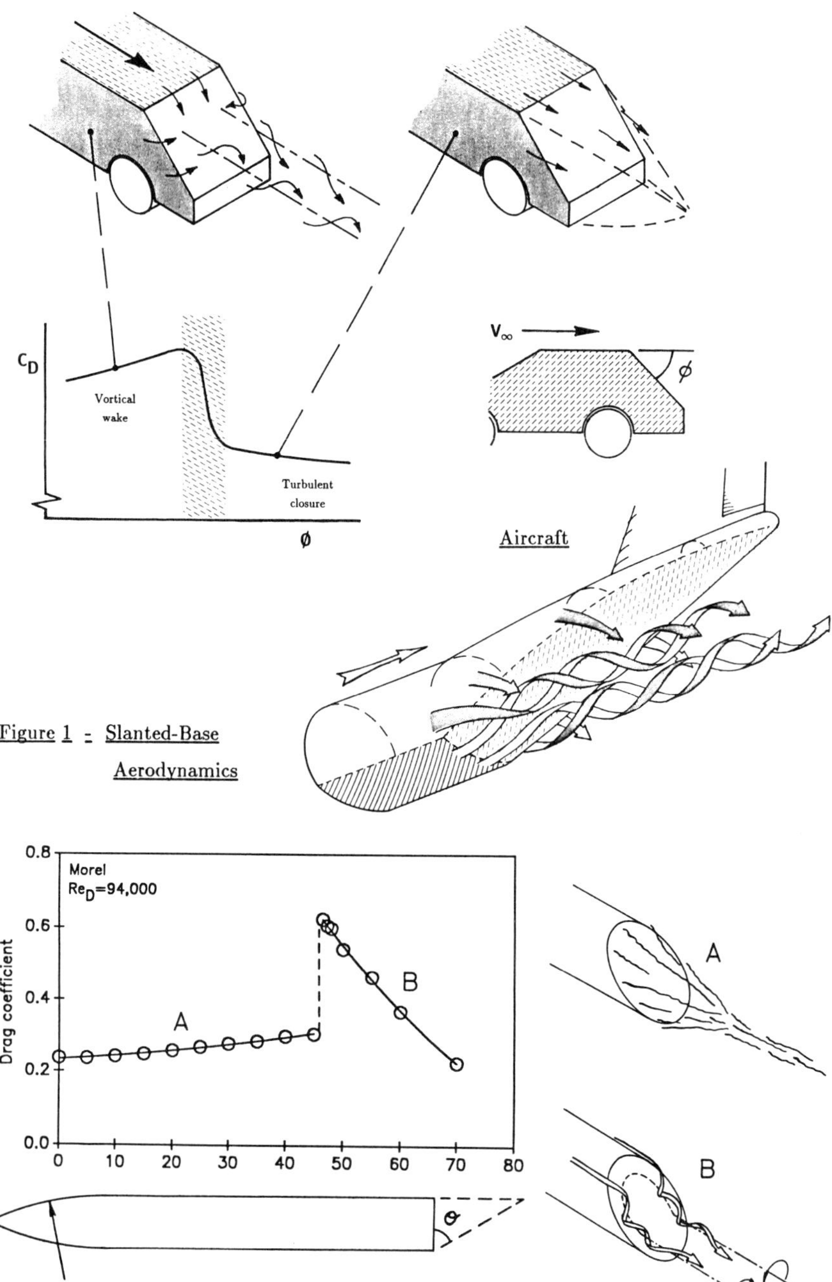

Figure 2 - Slanted-Base Ogive Cylinder Drag

angles to be tested. One model is used exclusively for force and moment tests, while an alternate, of nominally identical aerodynamic lines, is equipped for direct measurement of base pressures. For all tests, the models are suspended magnetically with no mechanical support of any kind. Previous experiments with these models have established their basic aerodynamic characteristics and are fully reported elsewhere[4,5,6].

The NASA Langley Research Center (LaRC) 13 inch Magnetic Suspension and Balance System (MSBS) has been developed from a system constructed at the Arnold Engineering Development Center in the 1960's[7], though little of the original hardware remains in use. The position and attitude of the model is detected optically[8], position signals are fed to a digital control system[9] with the electromagnet currents supplied by SCR power amplifiers. The wind tunnel is a low speed, open circuit design[10], illustrated in Figure 4, with a maximum Mach number of 0.5. An aluminum alloy dummy sting and support strut can be installed downstream.

A single channel onboard pressure telemetry system has been developed[6,11] to permit measurement of interference-free base pressures. Due to the low dynamic pressures encountered, this is quite a challenging measurement. With freestream static pressure and model base pressure both roughly atmospheric and with no direct connection between the model and the outside world, great care must be taken to overcome the "subtraction of elephants" problem. For these tests, this was accomplished by using a differential transducer with its backside vented to a total pressure tap in the extreme nose of the model. This tap is barely visible in Figure 3. Base pressures are derived from the difference between base static and tunnel total pressure, compared to the difference between local static and tunnel total.

CALIBRATION AND DATA REDUCTION

Previous analysis[6] has indicated that the drag and lift forces and pitching moment for a model at zero angles of attack and sideslip in the 13 inch MSBS can be expressed as follows:

$$F_X = a_1 I_L^2 + a_2 I_L I_D + a_3 I_D^2 + a_4 I_P I_L$$

$$F_Z = a_5 I_L^2 + a_6 I_L I_D + a_7 I_P^2 + a_8 I_P I_D$$

$$M = a_9 I_P I_L + a_{10} I_P I_D \tag{1}$$

where
$$I_L = I_1 + I_2 + I_3 + I_4 \quad ; \quad I_P = I_1 - I_2 + I_3 - I_4 \tag{2}$$

The numbering sequence for the electromagnet currents is shown in Figure 5. Coefficients a_1 through a_{10} are constants found by multiple regression fitting of calibration data. This data is obtained from a representative model (0° base), suspended and loaded at three stations.

Conventional corrections are applied to drag and pressure measurements to account for bouyancy and blockage effects[6]. Corrections for sting-induced buoyancy or blockage have not been applied, but would be extremely small. No corrections are made to lift forces or pitching moments.

All electromagnet currents can be measured to an accuracy of around ± 10mA (1mA instrument resolution). Measurements are an average of one hundred samples, taken under quasi steady-state conditions. This results in a typical uncertainty in drag force of less than 0.002N. The principal items of wind tunnel data are the total and dynamic pressures, measured with Datametrics 572D Barocells. Typical quoted accuracy is of the order of 0.2%, all sources. This will result in a Mach number uncertainty of the order of 0.1%. Additional errors accumulate due to innacuracies in measurement of stagnation temperature, imperfect calibrations and so forth, but are generally rather small.

Analysis of the effects of inaccurate estimates of a_1-a_{10} is very complex and is demanding of much further study. The correction of measured currents and predicted forces

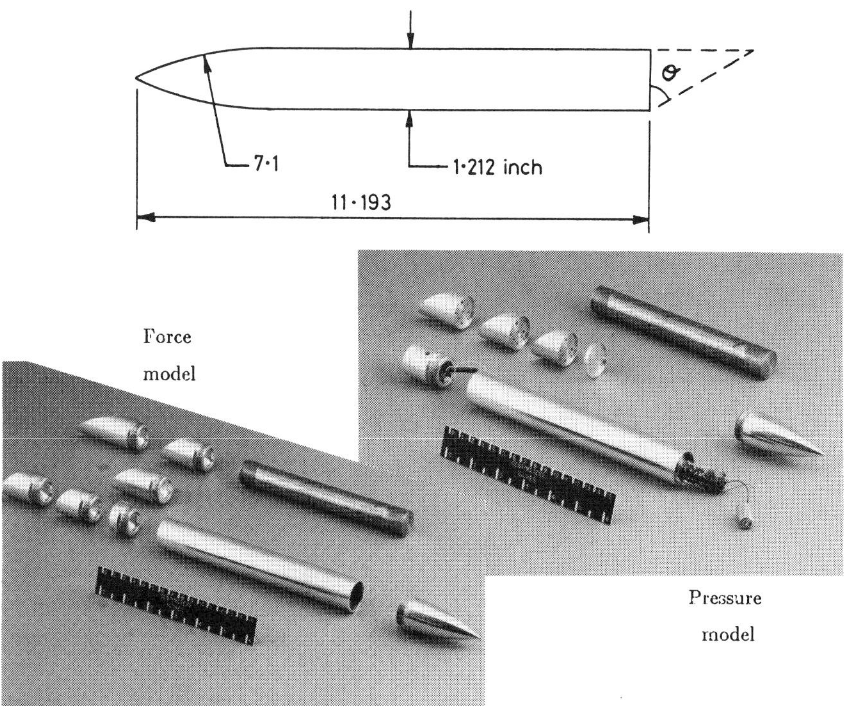

Figure 3 - Slanted Base Ogive-Cylinder Models for MSBS

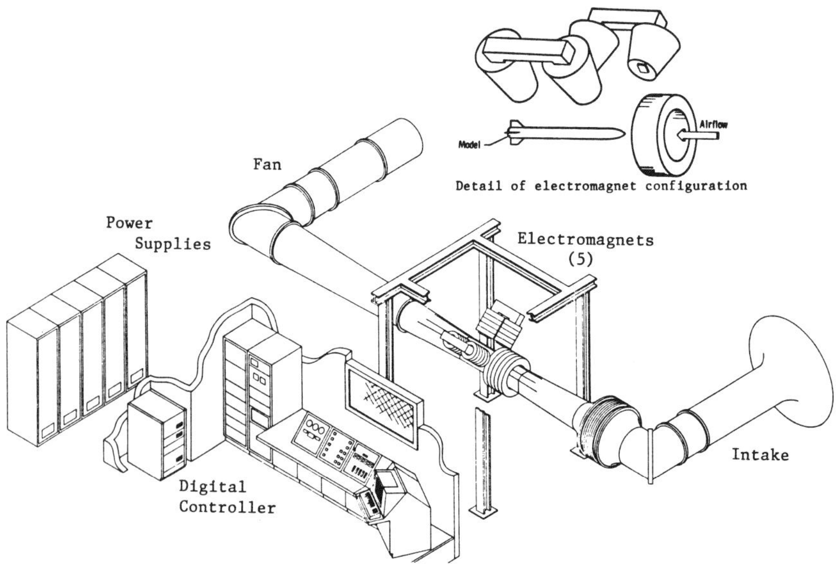

Figure 4 - NASA Langley 13 inch Magnetic Suspension and Balance System

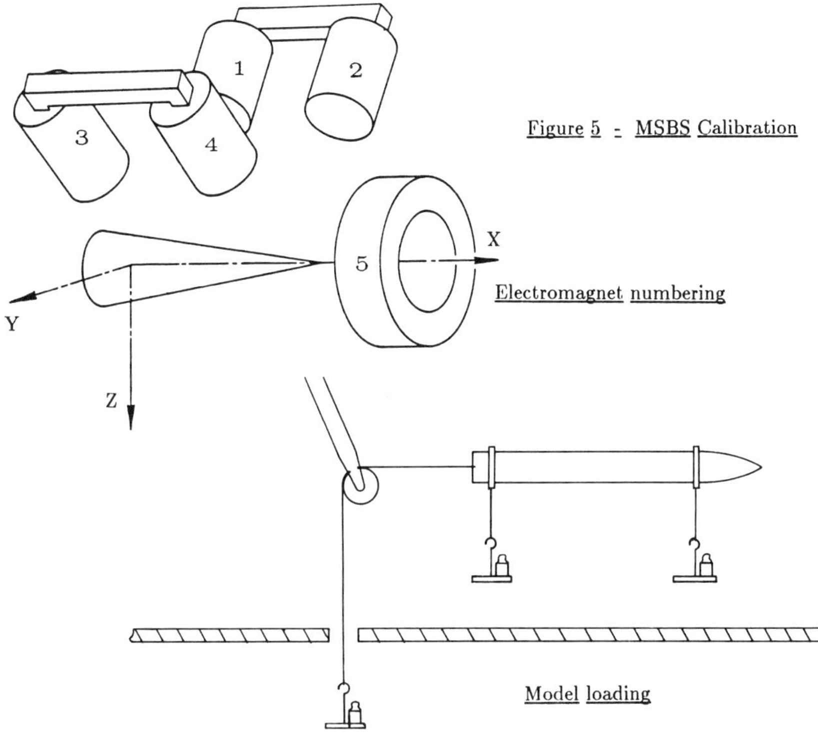

Figure 5 - MSBS Calibration

Electromagnet numbering

Model loading

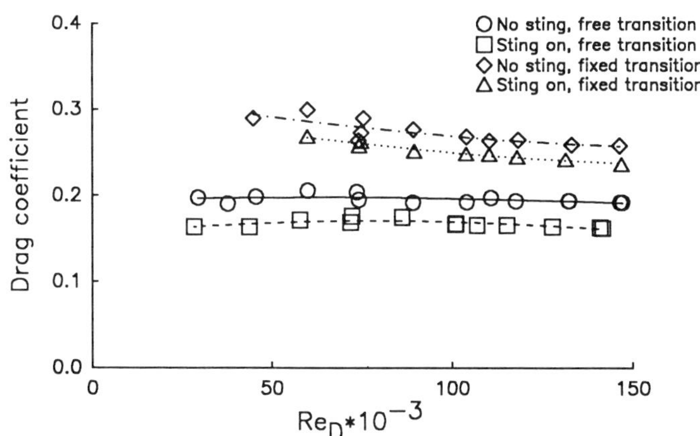

Figure 6 - Zero degree Base, Drag Coefficients

and moments by the wind-off zeros has the effect of cancelling some apparent error. Further analysis of the accuracy of results may be found in Reference 4.

AERODYNAMIC STING INTERFERENCE

Aerodynamic interference of the sting on the model may arise from various sources. Physical blockage downstream of the model introduces a pressure gradient into the flow in the region of the model. If the sting is at angle-of-attack or if the model's wake flow involves strong crossflow velocities (relative to the sting) then flow inclinations will also be induced in the region of the model. Boundary layer development, particularly on the aft regions of the model will be affected by the presence of the sting, partly due to the sting-induced perturbations. The wake geometry and structure are likely to be substantially affected, due to the physical presence of the sting within, or close to, the wake. The geometry of the aft region of the model is corrupted to accomodate the sting. This corruption includes, but is not limited to, truncation of the base and/or creation of a sting cavity.

Aerodynamic data must be adjusted or corrected for sting interferences. Repeat testing, with alternative sting arrangements and metric divisions of the model, is a viable and apparently the most common method of extracting the interferences. This approach can unfortunately lead to considerable complexity, with several test sequences required to extract the interference terms, due to the fact that no support system has zero overall interference. Analytic or empirical correction of test results is often attempted, though is fraught with its own difficulties.

For the tests reported herein, the geometrical distortion of the base is limited to the creation of a simple sting cavity. The sting is aligned with the tunnel axis and the models are non- or weakly lifting, so the overall flow disturbance caused by the sting is principally a weak longitudinal pressure gradient. The effects of the disturbances of the model's boundary layer and wake are thus highlighted.

EXPERIMENTAL RESULTS

Models were suspended close to the centerline of the wind tunnel, at nominally zero angles of attack and sideslip. The "clean" model is thought to exhibit principally laminar boundary layers, with transition just beginning to move forward from the base at the higher Reynolds numbers tested. Most tests were repeated with transition fixed 2cm downstream of the ogive-cylinder junction by a ring of No.60 grit. Some check runs were made with coarser grit to ensure tripping at lower Reynolds numbers.

Only a fraction of the available results are presented in this paper; more details can be found in Reference 4.

DRAG

Drag coefficients for the 0° base, with and without the dummy sting present, are shown in Figure 6. Slight reductions in drag coefficients with increasing Reynold's number is noticeable, particularly for the fixed transition cases, consistant with usual boundary layer behaviour. It is also clear that significant discrepancies exist between sting supported and interference-free data. The signs and magnitudes are, however, broadly consistent with previous measurements for ogive-cylinders, that is, somewhat lower drag with sting on (such as[12]). The mechanism for the drag reduction is thought to be the reduction in the required expansion of the flow surrounding the wake downstream of the base, since part of the flow area is occupied by the sting. This would result in slightly higher (less negative) base pressures in the sting-on case. Results for the 30° and 40° bases are broadly similar to the 0° case.

Previous MSBS tests[4,5,6] had revealed an unusual hysteretic behaviour of the wake for

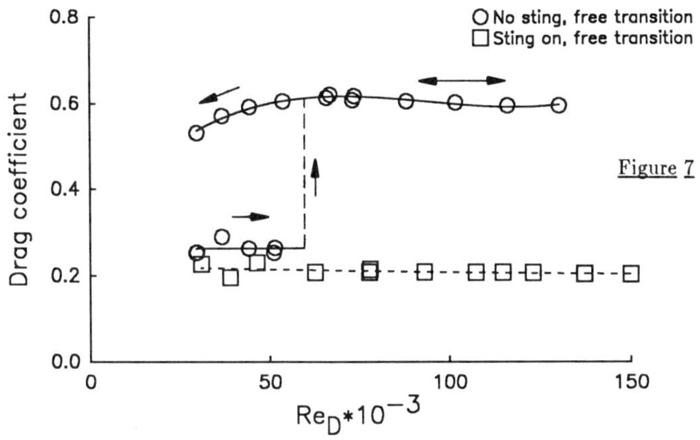

Figure 7 - 45 degree Base Drag Coefficients

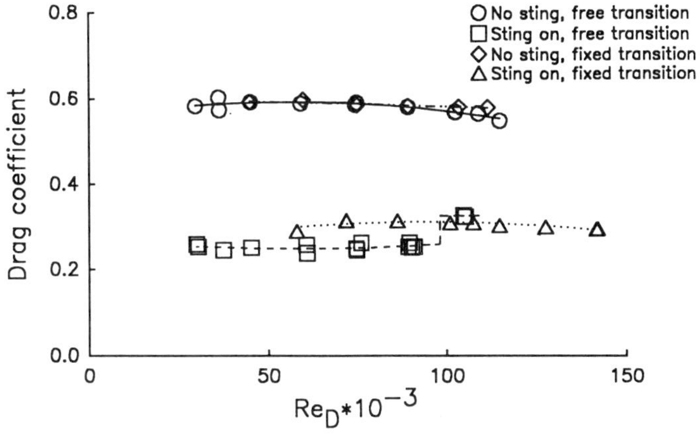

Figure 8

50 degree Base

Drag Coefficients

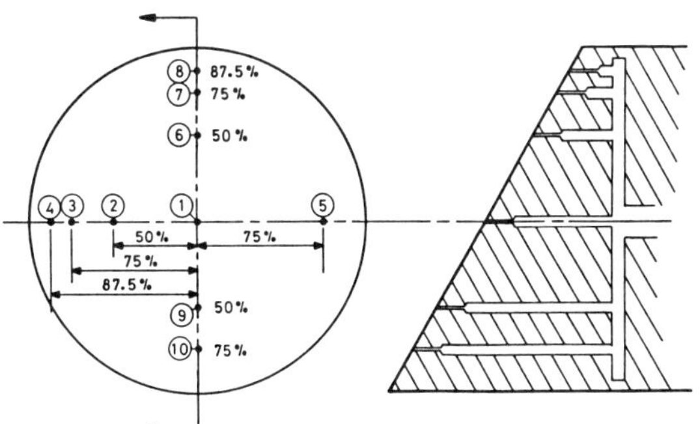

Figure 9 - Pressure Tap Pattern Schematic

the 45° base. As Re_D for the "clean" model increased past 60,000 or so, the wake structure suddenly changed from a quasi-symmetric pattern, characteristic of low slant angles, to a longitudinal vortex flow, characteristic of higher slant angles. The change in drag is dramatic, more than a doubling in value (Figure 7). Also, once the vortex flow is established, it persists even as Re_D is reduced below the "critical" value. It is believed, though not yet experimentally confirmed, that this change in wake structure is linked with the onset of natural transition at the model's base or in the free shear layers developing just downstream. Further, the fact that the effect had not previously been detected is thought to be due to boundary layer tripping by upstream wire or strut attachments[6].

It was found that tripping the boundary layer prevented this change in structure from occurring. Further testing revealed that the phenomena was sensitive to the grit size employed, since it is known that the No.60 grit normally used is too fine to ensure proper transition at the lowest Reynold's numbers. Repeat sweeps up and down through the "critical" Reynold's number with this size grit sometimes exhibited the wake flow change, sometimes not. A coarser grit size proved consistently effective in inhibiting the formation of the vortical wake. Not surprisingly, the introduction of a sting into the wake completely disrupts the changes in structure. With the sting present, the wake is always a quasi-symmetric closure, with no tendancy to form a vortical wake under any condition tested. The discrepancies between sting supported and interference free data thus become strong variables with large peak magnitudes. Drag is always lower with the sting present, due to the incorrect wake flow.

The interference-free wake for the 50° base slant is a steady longitudinal vortex flow, producing low base pressures, high drag and significant lift forces and pitching moments. At low Reynolds numbers with the sting present, the drag coefficient of the "clean" 50° base model remains fairly low, indicating a quasi-symmetric wake closure. At some critical Reynolds number, the wake changes spontaneously to a longitudinal vortex flow, with corresponding increase in drag, illustrated in Figure 8. No hysteresis has been detected. However, the drag increment in vastly less than that observed with the corresponding phenomena for the 45° base, interference-free. It can also be noted that the increment in drag between fixed and free transition, consistently observed for the lower slant angles, is not seen with the higher slant angles (interference-free). Since there must logically be an increment in forebody skin friction, suspicions are aroused as to whether the base drag changes by a similar increment, but in the opposite sense. This point is resolved shortly.

BASE AND STING CAVITY PRESSURES

The pressure tap pattern for base pressure measurements is illustrated in Figure 9. Base pressures as a function of Reynold's number for the 0° base are shown in Figure 10. No distinct trends are apparent. Sting cavity pressures are noticeably higher than base pressures, corresponding to the observed drag differences between the two cases (Figure 11). The reader is cautioned against comparing the _magnitudes_ of the pressure differences against the differences in drag, since the pressures on the annular region surrounding the sting cavity are not likely to be exactly equal to the cavity pressure.

The free transition base pressures for the 45° base, shown in Figure 12, dramatically illustrate the wake hysteresis mentioned earlier. The apparently steep variations of pressure coefficients with Reynold's number above the critical value are thought to be genuine, but may be due to slight shifts in the vortex locations, rather than actual changes in vortex strength or structure. Data below the critical Reynold's number should be considered unreliable, due to the very small pressure differences encountered. Figure 13 illustrates the distinctive pressure "footprint" of the vortices, with very high suctions in the region of the vortex cores. That tripping of the boundary layers inhibits formation of the vortical wake has been confined by pressure measurements.

Base pressures for the 50° base are shown in Figure 14. A reduction in the strength of

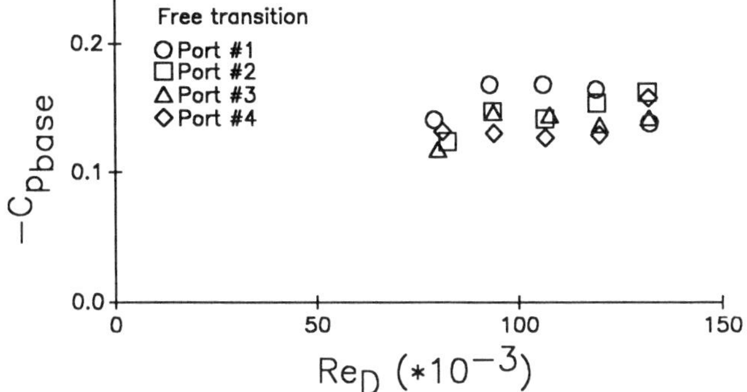

Figure 10 $\div$ Zero degree Base, Base Pressures

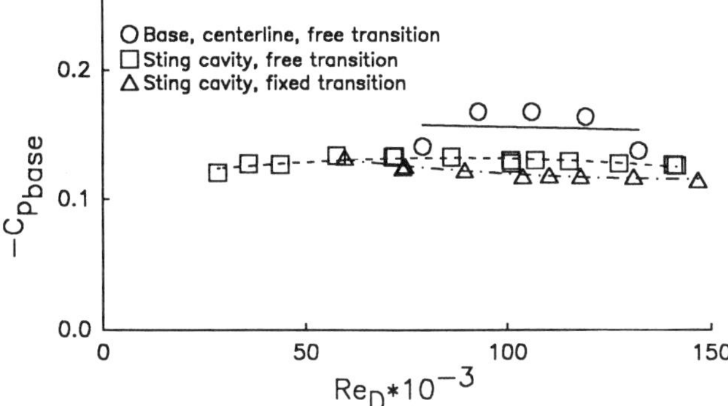

Figure 11 $\div$ Zero degree Base, Sting Cavity Pressures

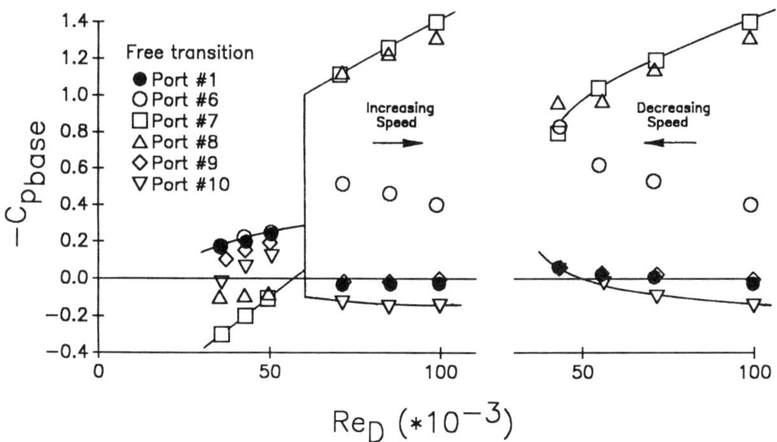

Figure 12 $\div$ 45 degree Base, Base Pressures

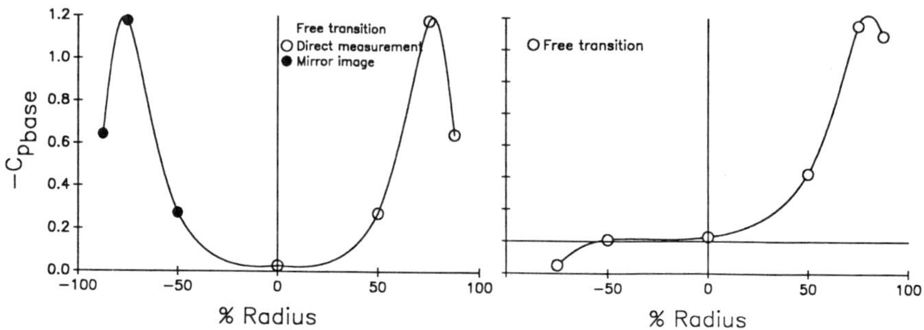

Figure 13 - 45 degree Base, Base Pressures, Re_D=94,000

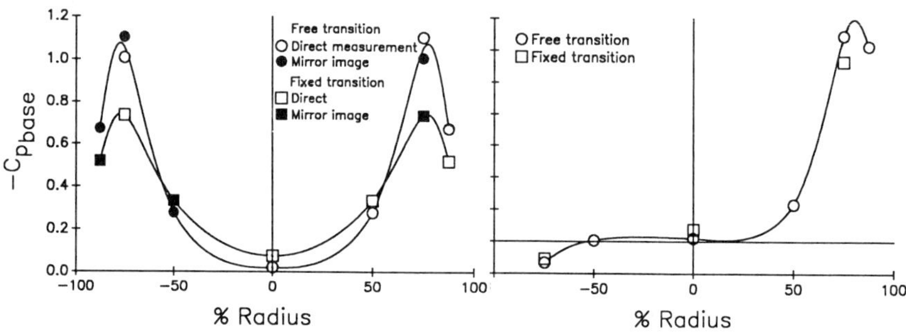

Figure 14 - 50 degree Base, Base Pressures, Re_D=94,000

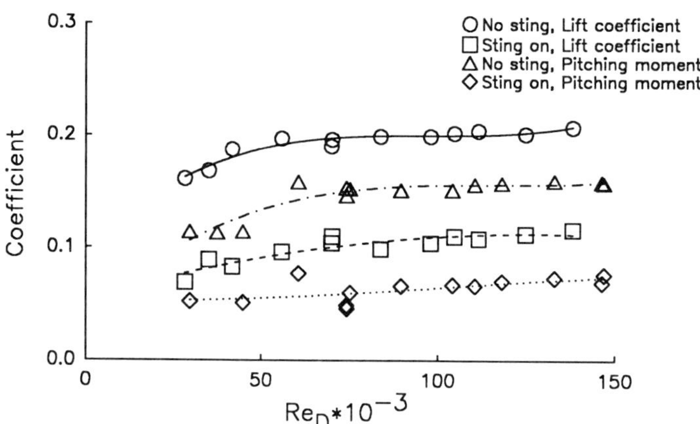

Figure 15 - 40 degree Base, Lift and Pitching Moment

the trailing vortices (corresponding to a reduction in base drag) with transition fixed was suspected and is clearly confirmed.

LIFT FORCES AND PITCHING MOMENTS

Lift forces are non-dimensionalized by the model cros-sectional area. Pitching moments are non-dimensionalized using the same reference area, also the half-length of the 0° base model. This choice of reference area results in apparently large numeric values of C_L and C_M, though the forces and moments involved are typically quite small. Results for the 40° base, shown in Figure 15, indicate relatively constant values of C_L and C_M, with consistantly lower values sting-on. This corresponds to the drag variations shown earlier, presuming that the majority of all forces and moments arise due to base pressures. Figure 16 again shows the characteristic hysteretic behaviour of the wake for the 45° base. Results for the 50° base, shown in Figure 17, also confirm the previously observed weakening of the wake vortices with transition fixed.

DISCUSSION OF RESULTS

Figure 18 shows a comparison of interference free drag coefficients to the original data by Morel ($Re_D = 94,000$[1]). MSBS points are obtained by interpolation for each base slant. Agreement for the interference free cases is quite good at the higher slant angles. At the lower slant angles, it is supposed that partial boundary layer tripping was induced by Morel's wire supports[6]. The critical slant angle was fractionally lower in the MSBS tests than had been previously reported. The apparent convergence of all results above the critical slant angle is again notable.

With sting-on, the drag is relatively constant for all base slants (Figure 19). For most angles the drag is too low, although at the highest angle tested the reverse is true. The data may appear somewhat scattered, but the reader is reminded that some form of vortex formation is quite likely around the 45° slant angle. Thus the points may lie on two or more distinct trend lines, as in the interference-free cases.

It is recognized that the use of a relatively massive sting support at subsonic speeds is rather unconventional. Thus the specific data presented herein might be regarded as unrepresentative of everyday interference problems. However, at transonic speeds, the use of some form of sting support is virtually universal and there is no reason to expect that interference phenomena should be confined to the subsonic regime.

MSBS IN A HELIUM FLOW FACILITY

The results presented above confirm that static aerodynamic testing of quasi-axisymmetric models may be accomplished in a MSBS. Multiple force and moment components can be evaluated simultaneously. This is despite the fact that the 13 inch MSBS is very poorly suited to this task, perhaps notably regarding lack of symmetry in the electromagnet configuration. Telemetry of pressure (or other) data from the suspended model is practical. Some observations particularly relevant to the helium flow application may now be made:

(1) The test section must be constructed of non-magnetic and preferably non-conducting material, in order to inhibit eddy currents. The test section of the 13 inch system is plexiglas and glass, noticeable in Figure 20. This might represent a significant design challange for a test section intended to "soak" at 4K or below, yet retain accurate geometry, optical access and so on.

(2) Calibration procedures are relatively elaborate and can be lengthy. To facilitate development and commissioning of the MSBS, it should be operable seperately from the helium tunnel, perhaps with a room temperature "test section" of some form.

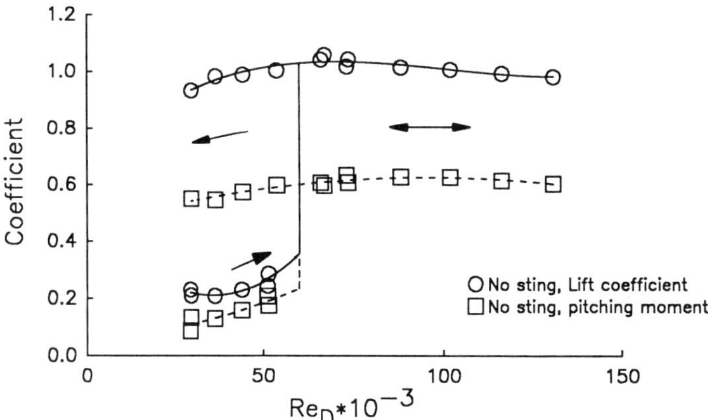

Figure 16 - 45 degree Base, Lift and Pitching Moment

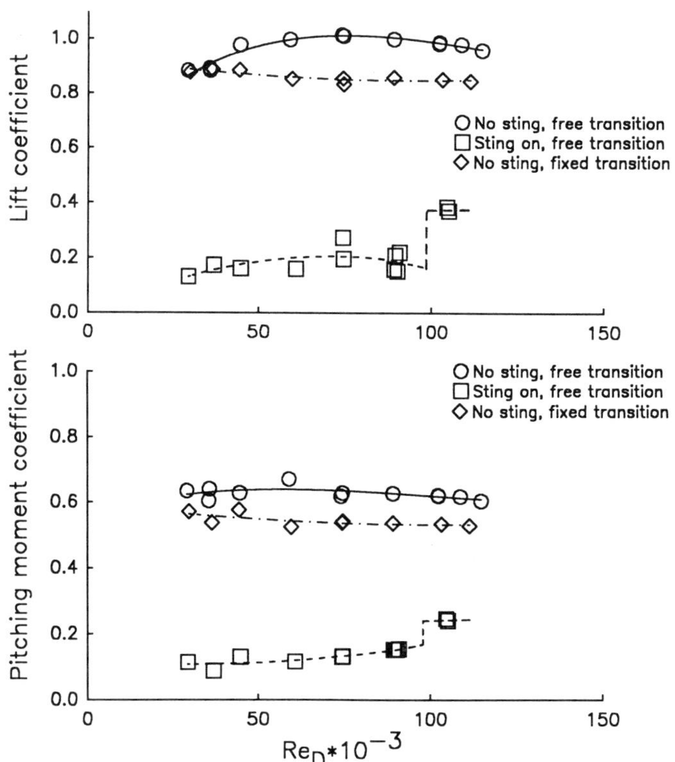

Figure 17 - 50 degree Base, Lift and Pitching Moment

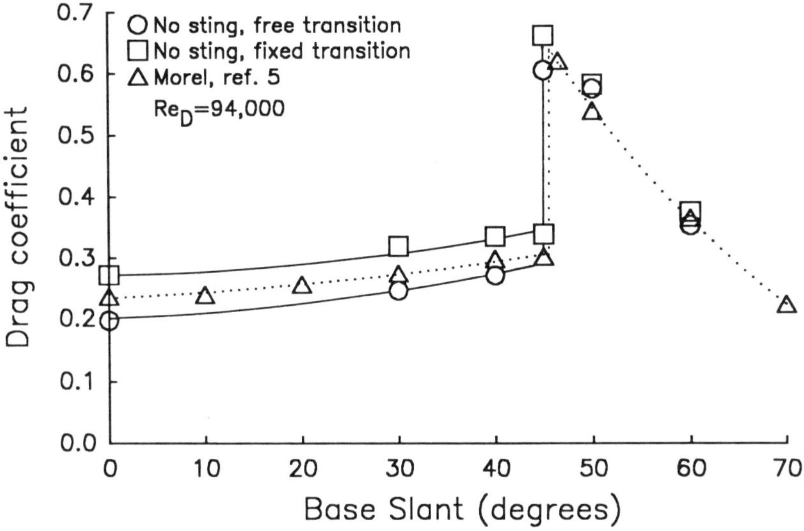

Figure 18 - Interference Free Drag Coefficients compared to Reference Data

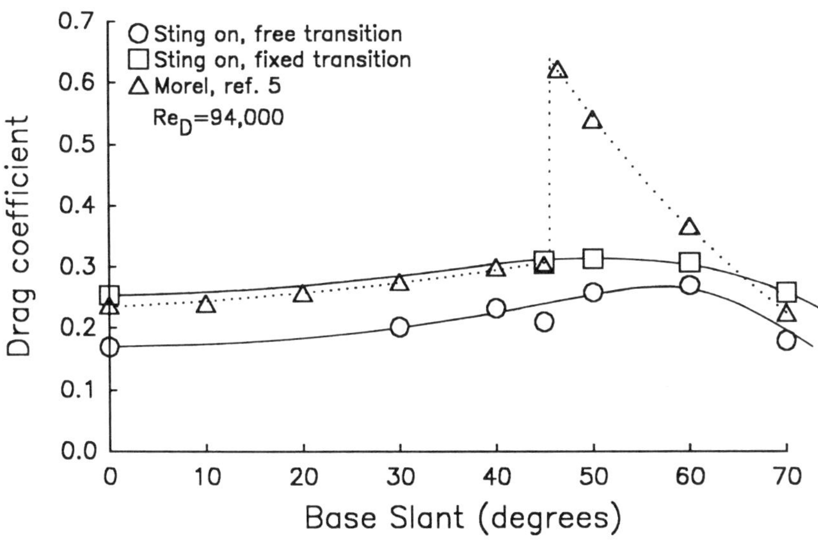

Figure 19 - Sting-On Drag Coefficients compared to Reference Data

Interference-Free

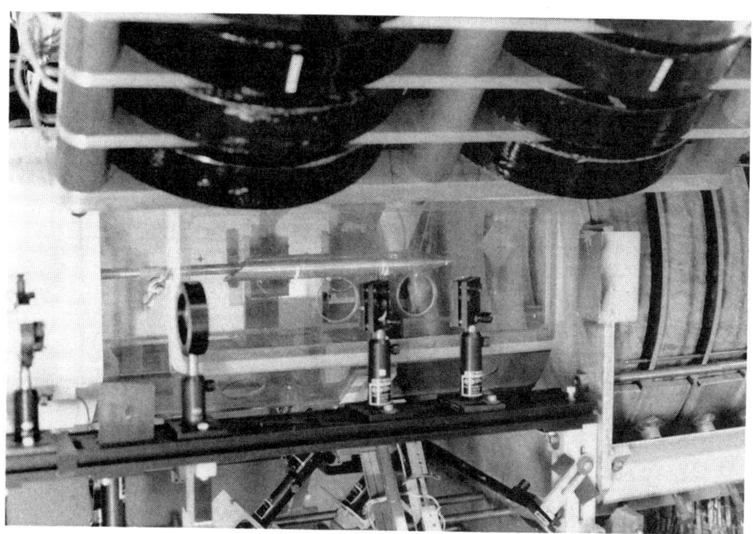

(Dummy) Sting-On

Figure 20 - Slanted-Base Ogive Cylinder Model in 13 inch MSBS

(3) Remote telemetry is undoubtedly practical (astronomers are already using electronic instrumentation cooled to 4K or so), but represents a significant development task.

(4) Three choices of model core are available; soft iron, as used in the 13 inch system, permanent magnets, used widely at Southampton University and superconducting solenoid cores, originally developed at Southampton[13]. The latter should give the highest force capability and seems attractive due to the "built-in" cooling provided by the helium tunnel. Other choices may be preferred for simplicity, however.

CONCLUSIONS

Sting interference is detectable for all slanted-base ogive-cylinder geometries and under all conditions tested. The magnitudes of the interference can be extremely large. Entirely different wake structures can exist with and without the sting present. Vortical wakes have been shown to be sensitive to the state of the oncoming boundary layer, with all aerodynamic parameters consequently affected. The 13 inch Magnetic Suspension and Balance System can be used successfully for measurements of multiple aerodynamic components.

REFERENCES

1. T. Morel, Symp. on Aero. Drag Mech. of Bluff Bodies and Road Veh., Plenum 1978, p.191.
2. X.J. Xia, P.W. Bearman, Aero. Qtrly., p.24 (1983).
3. D.J. Maull, Aero. Jrnl., p.164 (1980).
4. C.P. Britcher, C.W. Alcorn, W.A. Kilgore, NASA CR- (Sub. Jan. 1990).
5. C.W. Alcorn, C.P. Britcher, AIAA 15th Aero. Test. Conf., p.117 (1988).
6. C.W. Alcorn, NASA CR-181708 (1988).
7. R.P. Boyden, C.P. Britcher, P. Tcheng, SAE TP-851898 (1985).
8. P. Tcheng, T.D. Schott, ICIASF '87 Record, p.322 (1987).
9. C.P. Britcher, M.J. Goodyer, J. Eskins, D. Parker, R.J. Halford, ICIASF '87, p.334 (1987).
10. W.G. Johnson, D.A. Dress, NASA TM-4090, (1989).
11. P. Tcheng, T.D. Schott, 34th Int. Inst. Symp., p.407 (1988).
12. G. Lee, J.L. Summers, NACA RM-A57I09 (1957).
13. C.P. Britcher, M.J. Goodyer, R.G. Scurlock, Y.Y. Wu, Cryogenics, p.185 (1984).

ACKNOWLEDGEMENTS

This work was supported by NASA Langley Research Center under Grant NAG-1-716, Richmond P. Boyden, Technical Monitor.

Remarks on high-Reynolds-number turbulence experiments and facilities

K.R. Sreenivasan
Yale University

Abstract

A summary is given here of remarks made at the Workshop on the need for research in high-Reynolds-number turbulence and the possibility of using a helium tunnel for the purpose.

The main text

In an eloquent assessment that deserves to be read even today, von Neumann (1949) stated: 'The great importance of turbulence ... requires no further emphasis. ... Turbulence undoubtedly represents a central principle for many parts of physics, and a thorough understanding of its properties must be expected to lead to important advances in many fields. ... turbulence represents *per se* an important principle in physical theory and in pure mathematics. ...These considerations justify the view that a considerable ... effort towards a detailed understanding of the mechanism of turbulence is called for.' A few further remarks may be in order, especially espousing the need for experiments at high Reynolds numbers.

It has for long been clear that fluid turbulence is a significant engineering problem but, in spite of von Neumann's remarks, its place in mainstream physics has been ambiguous: turbulence has generally been thought to be too difficult and unfocused. The situation appears to be changing in recent years, and turbulence appears to be inching towards the important position in nonlinear physics that is its due.

The equations governing fluid turbulence have been known for long, but the information extracted from them has not been overwhelmingly large. To the extent that turbulent flows are nonlinear systems with many degrees of freedom which are strongly coupled, we have very little intuition about them. I cannot help but quote von Neumann again: ...our intuitive relationship to the subject is still too loose..., (and) we are still disoriented as to the relevant factors...' We are thus exploring new avenues all the time. The point I wish to make is that much progress in turbulence has eluded us at least partly because of the compromises that typical (that is, most) experimentalists are forced to make: Without proper tools, one cannot arrive at definitive facts. Without definitive facts, definitive theoretical ideas – which in turn feed the type of measurements that one should best make – cannot be developed. The atmosphere now seems right for this type of feedback.

I argue that several issues can be settled by first-rate experiments in high-Reynolds-number turbulence. Traditionally, most experimental research in universities has been carried out at moderate

Reynolds numbers, and the strong belief has been that the precise value of the flow Reynolds number is irrelevant as long as it is 'sufficiently high'. Most research wind tunnels thus cluster around a small region in Reynolds number. Many high-Reynolds-number experiments have been made in atmospheric and oceanic turbulence, and much has been learnt from them especially concerning the universality of small scale turbulence. The general feeling appears to be that a judicious combination of moderate-Reynolds-number experiments under controlled conditions and high-Reynolds-number experiments in geophysical flows serves our purpose for the most part. One might therefore wonder if there is indeed a compelling case to be made for expensive experiments at high Reynolds numbers. In other words, are there some specific and crucial issues to be settled for which one or more dedicated high-Reynolds-number research facilities are essential?

One might draw upon the few previous experiences when high Reynolds number facilities were made accessible to the turbulence community, albeit under sub-optimal conditions. For example, one has learnt something new and substantial from experiments such as those of Roshko (1961) and Kistler & Vrebalovich (1966), behind a cylinder and in grid turbulence respectively, made during the last days of the Southern California Cooperative Wind Tunnel before it was dismantled. Such focused measurements cannot be made in natural flows with uncontrollable experimental conditions and inherent non-stationarity. For example, it is impossible to make long-time measurements such as those required to determine high-order moments accurately. Indeed, there are suggestions that such moments diverge (Mandelbrot 1974, Schertzer & Lovejoy 1985).

One can also argue, simply from the point of view of providing hard data to which theoretical ideas have to be moored, that a well-executed experiment at high Reynolds numbers is worth more than many that are done under compromising conditions of Reynolds numbers, development length, resolution, etc. As in all asymptotic analyses, progress follows from an understanding of the high Reynolds number domain even if many (if not most) applications might pertain only to moderate Reynolds numbers. Yet, it is worth asking specifically: What critical issues need to be settled by new experiments at high Reynolds numbers? What intellectual and practical gains justify investing in them? Attempting an answer to these questions is too serious a task to be accomplished here, but a few points will nevertheless be made. They range (in arbitrary order) from the practical to the fundamental.

For almost fifty years now, Kolmogorov's (1941) ideas and their later modifications (Kolmogorov 1962, Obukhov 1962) have ruled the horizons of research in turbulence, and yet we are unclear about their status as a true theory. As an example, different theoretical ideas of recent origin predict that the celebrated $-5/3$ law is exact (e.g., Nakano 1986, 1988; Chorin 1988), approximate and needs a positive correction (e.g., Monin & Yaglom 1971), and approximate and needs a negative correction (Yakhot et al. 1989). Which of them is correct? What is the best way of experimentally settling this question? The situation is worse in sheared turbulence. How much of the structure one

sees in shear flows is a manifestation simply of the shear? How do the hairpin vortices oriented at 45°
to the mean flow direction survive at very high Reynolds numbers if the Kolmorogov cascade,
grinding inexorably towards the isotropy of small scales, is the principal dynamical mechanism? What
are the asymptotic properties of the probability densities of turbulent fluctuations? What is the nature of
turbulent mixing at very high Reynolds numbers? Do the small scales of a passive marker attain an
asymptotic state independent of the shear? What role does intermittency play in all these issues? The
uncertainties surrounding these questions are unsettling to students and professionals alike, unless
mere familiarity renders one immune to them.

Let me proceed along this line a bit further: What is the asymptotic value of the drag coefficient
for a high-aspect-ratio circular cylinder? What is the correct extrapolation formula for the skin-friction
coefficient in a two-dimensional boundary layer? Is the logarithmic region in the turbulent boundary
layer asymptotically consistent with the dynamics? Does the momentum transport in high-Reynolds-
number turbulent boundary layer occur principally in the form of intermittent events occupying
relatively small amounts of space and time? High-Reynolds-number experiments are especially critical
in the boundary layer because of the well-known difficulty of separating inner and outer scales at low
Reynolds numbers.

These (and other similar) questions can be narrowed down and perhaps resolved by means of
high-quality experiments at high Reynolds numbers. One way of attaining such high Reynolds
numbers is to build *substantially* bigger and better wind tunnels, water tunnels, pipe flows, or some
such standard facility. (An experience concerning a wind tunnel being built at IIT, Chicago was
discussed at the Workshop by H.M. Nagib.) A different suggestion is to build a helium tunnel whose
virtues were the subject of this Workshop. It is trivial to argue that a helium tunnel of *modest* size can
provide high enough Reynolds numbers that can be used for much of the standard model testing. The
Workshop revealed several instances where the feasibility and advantages of the helium tunnel have
been demonstrated. Much of the needed technology appears to be available. With such a facility one
can address several of the questions raised above (such as the drag on a flat plate and a circular
cylinder, or others such as decay rates of turbulence behind grids).

The situation, however, is unclear from the point of view of research in small-scale turbulence.
To decide on the feasibility of using helium tunnel for this purpose, one first has to address questions
such as: how can we resolve small scales of turbulence (e.g., vorticity, dissipation)? Can one obtain
meaningful spectral data? How much quantitative flow imaging will be possible? What are the intrinsic
limitations of resolution in experimental diagnostics? In short, how much new technology needs to be
developed before a helium tunnel becomes a good tool for turbulence research? These enquiries are
interesting in their own right.

Concluding remarks

I have been principally concerned in my remarks about turbulence research at high Reynolds numbers. I have briefly outlined some thoughts without the benefit of a thorough and quiet study. It is, however, clear that some thinking is required about the 'optimality' of the present mode of experimental research in turbulence. In this context, the potential of the helium tunnel should to be addressed adequately.

If one ignores the perspective of turbulence studies, it is clear that helium tunnels can extend the Reynolds number range accessible to most experimentalists. In particular, if they can be built without extraordinary expense so as to be accessible also to the university community, they acquire an added value.

References

Chorin, A.J. 1988 Comm. Math. Phys. **114**, 167.

Kistler, A.L. & Vrebalovich, T. 1966 J. Fluid Mech. **26**, 37.

Kolmogorov, A.N. 1941 C.R. Acad. Sci. USSR **30**, 301; **32**, 16. (For English translations, see *Turbulence: Classical Papers on Statistical Theory,* ed. S.K. Friedlander & L. Topper, Interscience: New York, 1961.)

Kolmogorov, A.N. 1962 J. Fluid Mech. **13**, 82.

Mandelbrot, B.B. 1974 J. Fluid Mech. **62**, 331.

Monin, A.S. & Yaglom, A.M. 1971 *Statistical Fluid Mech. Vol. 2.* M.I.T. Press.

Nakano, T. 1986 Prog. Theor. Phys. **75**, 1295; 1988 Prog. Theor. Phys. **79**, 569.

Obukhov, A.M. 1962 J. Fluid Mech. **13**, 77.

Roshko, A. 1961 J. Fluid Mech. **10**, 345.

von Neumann, J. 1949 In *Collected Works, Vol VI,* p.437 (ed. A.H. Taub, Pergamon Press, 1963)

Yakhot, V., She, Z.-S. & Orszag, S.A. 1989 Phys. Fluids **A1**, 289.

Helium

Micro and Macroturbulence in Superfluid Helium

K. W. Schwarz

IBM Research Division
Thomas J. Watson Research Center
Yorktown Heights, New York 10598

Abstract: The dynamics of superfluid helium are briefly reviewed. The application of this formalism to gain an understanding of the nature of microscopic (superfluid) turbulence is discussed. We point out that the problem of large-scale macroturbulence in the superfluid has remained largely untouched, and that it raises many interesting questions. In particular, it is not clear what basic equations one should use in discussing this problem, the usual Gorter-Mellink form being clearly inadequate. In fact, it is pointed out that the Gorter-Mellink approximation is of doubtful validity even in the situations where it has been traditionally applied. The main physical issues in formulating these kinds of problems are how the microturbulence interacts with the two velocity fields that describe the flow of superfluid helium to modify the macroscopic turbulence, and how this large scale turbulence acts back on and sustains the microturbulence.

I. Introduction

The prospect of a liquid helium wind tunnel raises interesting new possibilities for the study of fundamental fluid dynamical phenomena such as turbulence. By allowing operation both above and below the λ point at 2.12 K, such a facility will permit one to investigate the turbulent state in the normal (He I) and the superfluid (He II) regimes, and to compare the two. Neither of these states is particularly well understood, and by contrasting the two in the same experiment, new insights into the very important general phenomenon of turbulence may be attained.

The properties of ordinary liquids are familiar to the wind-tunnel community, but those of the superfluid state, if it is thought about at all, are considered to be too exotic to be worth bothering about. Closer examination, however, shows that the superfluid, too, is governed by a kind of classical fluid dynamics which may have important theoretical connections to such problems as the fluid dynamics of polymer solutions, liquid crystals, and other fluids which are permeated by an internal structure of some kind. This arises from the fact that two distinct kinds of turbulence can exist in He II. First, there is the kind of macroscopic, irregular, *chaotic* motion with which everyone is familiar, and which we will refer to here as *macroturbulence*. Although it can be expected to differ in many details, it is qualitatively similar to classical turbulence. It is also known, however, that in many situations He II can become microscopically turbulent, the entire fluid then being permeated with a dense tangle of quantized vortex-line singularities. This phenomenon of *microturbulence*, or *superfluid turbulence* involves entirely different scales from the motion associated with macroturbulence. Although the vortex singularities can weave their way over macroscopic distances, they have a core size which is only about an angstrom. Thus they resemble fine threads, many orders of magnitude finer than the Kolmogorov cutoff scale for classical turbulence, which fill the superfluid and which interact with and modify the macroscopic flow. The analogy to a dilute solution of long polymers is obvious, although the implications of this analogy have not yet been explored.

Because it represents an entirely new physical phenomenon associated with the superfluid state, microturbulence has been very extensively studied by low temperature physicists.[1,2] As will be recapitulated below, it is by now probably better understood than ordinary turbulence. In contrast, almost no fundamental work has been done on the nature of macroscopic turbulence in He II, and in particular on the question of how the large-scale turbulent motion couples to and is modified by the microscopic vortex tangle. Strong effects can be expected, and much work will be needed to sort things out. Furthermore, although superfluid helium may be an esoteric substance, its behavior is not without engineering interest. It is an important cryogenic coolant for superconducting magnets, and it plays an increasingly important role in outer-space applications. Thus from this more practical point of view it is also of interest to learn more about its flow properties.

II. Helium II as a Classical Liquid

One does not need to understand the molecular theory of fluids in order to do fluid dynamics: a classical dynamical description on the order of the Navier-Stokes equation is sufficient. Just so, it is possible to use a "classical" dynamical description of the flow properties of superfluid helium without understanding the complicated and still only imperfectly realized quantum mechanical theory of the superfluid state. Because the underlying microscopic nature of the superfluid differs greatly from that of an ordinary fluid, these dynamical equations are of course not the same as those which govern ordinary liquids, but this need not intimidate us: we know of many systems, e.g. plasmas, liquid crystals, polymer solutions, and so on, which require one to use something more than the Navier-Stokes to describe them. In this spirit, we can think of He II as a classical fluid having certain peculiar additional features arising from the special nature of the fluid.

Let us first consider the notion of He II at absolute zero, when no thermal excitations are present. The first peculiar feature of the superfluid is that it has no viscosity. Rather than the Navier-Stokes equation, it obeys the Euler equation[3]

$$\rho \frac{D\vec{v}_s}{Dt} = -\nabla p \quad , \tag{1}$$

where D denotes the hydrodynamic derivative. A second peculiar feature is that superflow must always be irrotational -- there is no such thing as local or global solid-body rotation in He II. This alone serves to guarantee that macroturbulence in He II must, on some level, differ significantly from classical turbulence which is known to involve rotational eddies. The irrotational condition is expressed by

$$\nabla \times \vec{v}_s = 0 \quad . \tag{2}$$

As in an ordinary fluid, one can assume that fluid to be incompressible. Hence,

$$\nabla \cdot \vec{v}_s = 0 \quad . \tag{3}$$

The extreme simplicity of this description seems to good to be true. In fact, the irrotational, incompressible, frictionless flow described by Eqs. (1) - (3) can be observed in the laboratory without difficulty. Thus, if one proceeds with sufficient delicacy, it is possible to set up a flow in a toroidal channel, and watch it circulate without slowing down, for as long as one wants to. Nevertheless, our suspicions are seen to be justified if we try to increase the flow velocity -- a *critical velocity* is soon reached at which something new happens. That new thing is the appearance of vortex filaments (Fig. 1) in the flow filed $\vec{v}_s$. Note that Eq. (2) is violated in the core of such a filament. However, the core radius a_0 of these vortex filaments is fixed by the microscopic theory at about one angstrom, so that the core region plays no significant role in the fluid dynamics. The microscopic nature of the He II also fixes the circulation κ of these vortices:

$$\kappa = \oint \vec{v}_s \cdot \vec{d\ell} = h/m \quad , \tag{4}$$

where h is Planck's constant and m is the mass of the helium atom. The existence of these *quantized vortices*, the behavior of which is that of a classical vortex filament, but whose parameters are determined by quantum mechanics, is a third peculiar feature of the superfluid state.[4]

When the critical velocity is reached by the superfluid flowing through a channel, the flow field suddenly fills up with a tangle of these quantized vortices. The line length density of such a tangle can vary from a few centimeters per cm^3 to thousands of kilometers per cm^3. Although this is clearly very different from classical turbulence, the appearance of such a constantly evolving, self-sustaining quantized-vortex tangle does represent a transition of the flow from a laminar to a chaotic state, a state which has traditionally been called superfluid or quantum turbulence, and which we here also refer to as microturbulence to emphasize the fact that it does not involve large-scale turbulent motions. Rather, it is perhaps best thought of as a dynamical modification of the internal structure of the fluid, which is now permeated by these microscopic whirlpools.

The properties of superfluid turbulence can be calculated. Before turning to this, however, we need to mention one final important peculiarity of the superfluid state. At finite temperatures, the liquid will contain a gas of thermal excitations (vibrations). In an ordinary fluid, thermal excitations scatter so strongly that they hardly exist as identifiable entities. In He II, by contrast, excitations propagate unrestrained, interacting only with each other and with boundaries, much like particles is a gas. The end result is that, at T > 0, He II contains a *normal fluid* of thermal excitations,[3] which acts like a classical fluid and flows without friction relative to the superfluid velocity field $\vec{v}_s$. The fraction of the total fluid density associated with this normal fluid velocity field $\vec{v}_n$ increases as the temperature is raised, reaching unity at the λ point. Since this motion carries with it the entire heat content of the fluid, the fluid dynamical behavior described by $\vec{v}_n$ is inseparable from the heat flow behavior. This leads to various spectacular effects such as enormous rates of heat transfer (normal fluid flow $\vec{v}_n$) as well as flow at absolute zero (superfluid flow $\vec{v}_s$).

When quantized vortices are present, the picture becomes more complicated. The thermal excitations are now scattered by the microscopic vortex singularities, so that the normal fluid drifting past a vortex line exerts a frictional force on the superfluid in the neighborhood of the core. Thus, as the normal fluid blows through the vortex tangle, a *mutual friction* body force is generated which couples the normal fluid motion and the superfluid field, and this greatly modifies the overall behavior. In particular, $\vec{v}_s$ now becomes dissipative, since it can exchange energy and momentum with the normal fluid and hence the environment. Also, $\vec{v}_s$ and $\vec{v}_n$ now have a strong tendency to come to the same value, since the mutual friction force coupling them together can easily become much larger that the viscous forces acting in the normal fluid.

III. Superfluid Turbulence

To understand the microturbulent state, we consider the situation in Fig. 2. Suppose that one has laminar normal and superfluid driving fields, characterized by the constant velocities V_n and V_s. To determine whether this simple flow is stable against the appearance of microturbulence, we add to this flow some initial configuration of vortex lines, and then determine how this tangle evolves in time. The idea is to perturb the flow in allowed way and, by allowing the vortices to evolve according to the laws of fluid mechanics, do the equivalent of a nonlinear stability calculation. Does the perturbation decay away, or does it grow to fill the entire channel with a self-sustaining vortex tangle? This project may seem somewhat ambitious, since it requires integrating something as complicated as a vortex tangle forward in time. However, sufficient simplifications are possible to make this a viable approach.

The first step in such a program is to realize that the velocity field $\vec{v}_s(\vec{r}, t)$ generated by a configuration of vortex singularities, such as those shown in Fig. 2, can be written down immediately in terms of the Biot-Savart law, since the vortices act as sources for the $\vec{v}_s$ field in exactly the same way as a current-carrying wire acts as a source for a magnetic field. To this field we can add the driving field $\vec{V}_s$ and any corrections arising from the presence of boundaries. In an ideal fluid, however, a vortex must move with the fluid, a rule known as the Kelvin circulation theorem. Hence the field we have just calculated from the vortex configuration tells us just how the configuration is to be moved. One can teach a computer to do this -- calculating $\vec{v}_s$ everywhere on the vortex lines, moving the lines around accordingly, recomputing $\vec{v}_s$, and so on. At finite temperatures, the effect of the normal fluid friction on this motion must also be considered. It turns out to enter as a relatively simple correction to the motion that the vortices induce on themselves. The upshot of applying the sequence of steps outlined above is that every point on the vortex tangle moves with a velocity given by[5]

$$\dot{\vec{s}} = \vec{V}_s + \beta \vec{s}\,' \times \vec{s}\,'' + \alpha(T)\,\vec{s}\,' \times (\vec{V}_n - \vec{V}_s - \beta \vec{s}\,' \times \vec{s}\,'') \ . \tag{5}$$

Here $\vec{s}(\xi, t)$ is the function describing the location of the vortex singularity in terms of some parameter ξ, e.g. the arc length, as well as the time t. The primes denote differentiation with respect to the arc length, so that the terms $\vec{s}'$, $\vec{s}''$, and $\vec{s}' \times \vec{s}''$ have the interpretation shown in Fig. 3, $\beta = (\kappa/4\pi) \ln(s''a_0)$ is a scaling parameter, and $\alpha(T)$ is a friction constant which describes the interaction between the normal fluid and the vortex singularities.

The first term in Eq. (5) shows that the vortices are flushed along with the mean superfluid driving velocity in the channel. The $\beta\vec{s}' \times \vec{s}''$ term is the motion that a vortex filament generates on itself because it is curved. This term is large here because of the almost singular nature of the the vortex core, and dominates any long-range contributions that arise from integrating the Biot-Savart equation over the rest of the vortex tangle. According to this term, every part of the vortex filament moves along its local binormal as shown in Fig. 3, at a rate inversely proportional to the local radius of curvature. For a random configuration, such as that shown in Fig. 2, this implies a constant, complicated three-dimensional internal churning of the vortices in the tangle. The final term in Eq. (5) describes the effect of the normal fluid friction acting on the vortex cores. Note that the quantity in parentheses is the relative velocity between the normal fluid and the vortex. This term is important because it can cause vortex lines to grow or decay, as shown in Fig. 4.

Equation (5) can be applied to an initial vortex configuration to see how it develops.[6]   One important feature, however, must be added before everything makes sense. During the random internal motion of the tangle, the situation frequently arises that one vortex tries to cross another (Fig. 5). We deal with this situation by allowing the vortices to reconnect as shown. Note the very important fact that this process allows vortex singularities to multiply in number, thus making possible the vortex population explosions which signal the development of the superfluid state.

Fig. 6 shows the time development of an initial set of vortices, computed according to Eq. (4) and the reconnection assumption. One finds that the vortex tangle reaches a certain limiting density about which it then fluctuates. Amazingly enough, the limiting line length density calculated from this simple fluid dynamical model agrees very well with that measured in real superfluid turbulence. This can be seen from Fig. 7, for example, which compares the mutual friction force measured in a particular experiment with that calculated for the vortex tangle on the computer.

IV. Speculations on Macroturbulence in HeII

Although there are many other questions that one can ask about the nature of superfluid turbulence, the results discussed in the previous section make one confident that one has a pretty good idea of what is going on physically in the vortex tangle. Can one make use of this understanding to develop modified macroscopic equations, generalizations of the usual two-fluid equations which include the effects of the microscopic vortex-tangle in the form of extra terms? Can one guess at the effect that the internal vortex-tangle structure would have on ordinary macroscopic turbulence, and would one even expect the microturbulent and macroturbulent states to coexist? What is fascinating about these kinds of questions, and what makes further experimental studies of the kind made possible by the proposed facility so important, is that at present we have no real insight into what the answers are likely to be. One can at most point to some particularly important issues and offer some very speculative suggestions.

Because of the two-fluid nature of He II, there are two fundamentally different types of situation to consider. First, there is the common situation that there is a large relative velocity between $\vec{v}_n$ and $\vec{v}_s$. Because $\vec{v}_n$ carries the heat, this case is one of large thermal transport, such as might arise in cooling applications. Such a situation also obtains, for example, in the operation of fountain pumps, which are currently under intensive development as a means of liquid helium transfer in outer-space applications. From the basic-research point of view, it is known that this situation is precisely what is needed to support intense microturbulence. By having $\vec{v}_n$ and $\vec{v}_s$ in opposite directions (counterflow), one can arrange to have little or no mass flow, and hence at the same time minimize the likelihood of macroscopic turbulence.

A second extreme is to have large mass flow, with little or no relative velocity between the normal fluid and the superfluid. Such a situation can be expected to occur if He II is made to flow by some mechanical means, such as an impeller pump. Since it is $\vec{v}_n - \vec{v}_s$ which drives the microturbulence, it is expected that microturbulence does not occur, or at least is greatly reduced in this limit. On the other hand, there is every reason for expecting ordinary (macro) turbulence in such flows. A certain folklore has developed about each of these situations, and our purpose here is simply to point out some problems with the conventional wisdom.

We first consider the case where the flow is primarily counterflow (Fig. 2). Strong microturbulence will be present and the first question is how to modify the equations to account for this internal structure. The recipe used for doing this is essentially unchanged today from the time when the problem was first considered -- one simply adds a mutual friction term $\vec{F}_{sn}$, which is supposed to describe the effect of the interaction between the vortex tangle and the normal fluid, to the acceleration equations of the normal and superfluids

$$\rho_n \frac{D\vec{v}_n}{Dt} = -\frac{\rho_n}{\rho} \nabla p + \eta \nabla^2 \vec{v}_n - \rho_s \nabla T - \vec{F}_{sn} \quad , \tag{6}$$

$$\rho_s \frac{D\vec{v}_s}{Dt} = -\frac{\rho_s}{\rho} \nabla p + \rho_s s \nabla T + \vec{F}_{sn} \quad . \tag{7}$$

Of course, one needs now to express $\vec{F}_{sn}$ in terms of $\vec{v}_n - \vec{v}_s$ itself. This can be done (as it was first done in 1949 by Gorter and Mellink[8]) on crude phenomenological grounds, or by using the theory described in the previous section. In either case, one writes an approximate relation

$$\vec{F}_{sn} = a(T)(v_n - v_s)^2(\vec{v}_n - \vec{v}_s) \quad , \tag{8}$$

where a(T) is a known function. This equation reflects the fact that the density of line in the vortex tangle is proportional to the relative velocity squared, while the friction between the vortices and the normal fluid is linear in the relative velocity. Equations (6) - (8) form the mainstay of engineering analysis of things like fountain pumps, and have been accepted without question also in the low temperature physics literature.

Yet, on some level, we can see now that these equations are clearly wrong. The question is what one means by $\vec{v}_s$ in this situation. It is not what we meant earlier, in Eqs. (1) - (5), where $\vec{v}_s$ was required to be irrotational except on the cores of the quantized vortex singularities. Clearly, we are now thinking of $\vec{v}_s$ as some macroscopic variable, obtained by averaging over the microturbulence. So far, so good: for the situation shown in Fig. 2, $\vec{v}_s$ would then just be the constant local driving velocity $\vec{V}_s$, the tangle being averaged away. That is, there is no inconsistency in applying Eqs. (6) - (8) to the case where $\vec{v}_n$, $\vec{v}_s$, and $\vec{F}_{sn}$ are everywhere steady and constant, since this is just the idealization which, by the results of the previous section, gives rise to Eq. (8). Under any conceivable realistic scenario, however, $\vec{v}_s$, $\vec{v}_n$, and $\vec{F}_{sn}$ will vary across the channel. At the least, $\vec{v}_n$ has to go to zero at the walls, and this in turn will affect the other quantities, if only near the walls. Thus any attempt to apply realistic boundary conditions to these equations implies that $\vec{v}_n$, $\vec{v}_s$, and $\vec{F}_{ns}$ be treated as spatially varying quantities.

There are two problems with this. For one thing, Eq. (8) is understood to apply specifically to the case of spatially homogeneous microturbulence. The solution we obtain, however, shows that it cannot be so since the driving fields are not homogeneous. This in all likelihood is not a major problem, since one can argue that Eq. (8) could be a good local approximation even when the global solution is not strictly constant. More disturbing is the fact that the macroscopically defined $\vec{v}_s$ found in this situation is no longer irrotational. The only way one can locally average an irrotational field containing vortex lines to obtain a macroscopic field with a nonzero curl is if the vortex lines are not randomly arranged. That is, only if there is some preferential alignment of the vortex lines along particular directions can one obtain a spatially varying averaged $\vec{v}_s$. This notion, however, is entirely at odds with the approximations leading to Eq. (8) in the first place. At the very least, it does not make sense to conclude that the vortex tangle has large scale correlations in its internal structure because of an equation derived assuming such correlations are absent. Yet that is the conundrum we face in applying the Gorter-Mellink approximation.

To resolve this puzzle conceptually, and perhaps to come up with a more refined version of the Gorter-Mellink equation, it would seem useful to extend the vortex tangle calculations described in the previous section to the case of nonuniform driving velocities. For example, one might assume a given $\vec{v}_n$ distribution and then compute the vortex-tangle structure and the macroscopic average field $\vec{v}_s$ self-consistently. Projects like this, which have not yet been attempted, could provide useful information on the extent to which large-scale correlated structures must exist in the quantized vortex distribution, and on the extent to which these can affect the macroscopic dynamics and require modification of the Gorter-Mellink approximation.

Finally, we turn to the second prototypical case, that of pure mass flow. To the author's knowledge, the occurrence and the characteristics of macroturbulence in this situation has not been a concern of the low temperature physicists who have traditionally studied the dynamics of He II. Thus there is very little experimental information, or, perhaps more accurately, very few observations which have been interpreted in terms of macroscopic turbulence. Because of its engineering interest, however, some rather impressive large-scale studies at very high flow rates are now being carried out (see the artice by Van Sciver in this book), and these promise to provide experimental information that may also be of interest from the basic research point of view.

Nothing whatever exists on the theoretical side of the subject, not even an approximation on the level of the Gorter-Mellink equations. One might in fact take Eqs. (6) - (8) as the starting point for macroscopic turbulence in He II, in the same way as the Navier-Stokes equation is the basic equation for describing turbulence in a classical fluid. This would be equivalent to assuming that the vortex tangle always reaches local steady-state equilibrium with respect to the local driving fields $\vec{v}_n$ and $\vec{v}_s$. Turbulent velocity fields fluctuate, however, and the response of the vortex tangle to changes in the driving velocities is known to be rather slow. Hence this approximation, which locks the local line-length density to $\vec{v}_n$, $\vec{v}_s$, is not likely to be sufficient. This suggests a more general approach in which the line-length density $L(\vec{r}, t)$ is considered as a third field to be calculated, along with $\vec{v}_n$ and $\vec{v}_s$. The relation (8) is now replaced by an equation

$$\vec{F}_{sn} = b(T)L(\vec{v}_n - \vec{v}_s) \ , \tag{9}$$

relating $\vec{F}_{ns}$ to the line length density, and an *additional* equation, first derived by Vinen,[1]

$$\frac{\partial L}{\partial t} = c_1 |\vec{v}_n - \vec{v}_s| L^{3/2} - c_2 L^2 \ , \tag{10}$$

describing the transient behavior of the vortex tangle. This results in a complete set of equations at the level of the Gorter-Mellink approximation, and could form a basis for investigating the macroturbulence problem. Of course, these equations are open to the same objections as before concerning the possible importance of correlations in the vortex tangle.

A much simpler viewpoint has been expressed several times at this conference: that, at least for very high flow rates, macroturbulence in He II is indistiguishable from turbulence in ordinary helium. The experimental observations of Van Sciver seem to support this simple-minded point of view. One might argue that microturbulence can be expected to be present and that the mutual friction due to the internal vortex structure will then couple the $\vec{v}_n$ and $\vec{v}_s$ fields together so strongly that they move as one fluid. This may be a useful engineering approximation, but it raises many unanswered questions. For one thing, if $\vec{v}_n$ and $\vec{v}_s$ are really locked together, there is nothing to support the microturbulence which depends on a difference between these velocities to keep it alive. It would seem more realistic to expect that some compromise must be reached involving a sufficient mismatch between the fluctuations in $\vec{v}_n$ and $\vec{v}_s$ to keep the vortex tangle at a high density. The notion that $\vec{v}_n$ and $\vec{v}_s$ are perfectly coupled must also break down near boundaries, since there $\vec{v}_n$ is forced to go to zero. This may be more important at lower flow rates, in the regime where macroturbulence first arises. Similarly, even if He II closely resembles an ordinary fluid for the pure mass-flow case, it does not do so when there are large heat currents. As heat flow becomes important, therefore, the notion of perfect coupling must certainly become incorrect, and it is interesting to consider what would happen to the macroturbulence in this crossover regime.

V. Conclusion

The main message of this paper has been that there are many major unanswered questions when one confronts the issue of macroscopic turbulence in He II, questions which could be investigated using a large, high Reynolds number facility such as the proposed wind tunnel, where one could go from the normal to the superfluid state. In particular, how large scale macroturbulence is affected by the rather well-understood phenomenon of microturbulence, and how it compares to turbulence in He I are completely open questions. Both from the fundamental physics viewpoint and from an engineering perspective, these are important issues to be addressed in the future.

References

1. W. F. Vinen, Proc. R. Soc. London, Ser. A **240**, 114 (1957); **240**, 128 (1957); **242**, 493 (1957); **243**, 400 (1957).

2. J.T. Tough, in *Progress in Low Temperature Physics*, edited by D. F. Brewer (North-Holland, Amsterdam, 1982), Vol VIII, p. 133.

3. L.D. Landau and E.M. Lifshitz, *Fluid Mechanics*, (Pergamon, London, 1959) p.507.

4. R.P. Feynman, in *Progress in Low Temperature Physics*, edited by C.J. Gorter (North-Holland, Amsterdam, 1955), Vol.I, p.15.

5. K.W. Schwarz, Phys. Rev. B **31**, 5782 (1985).

6. K.W. Schwarz, Phys. Rev. B **38**, 2398 (1988).

7. D.F. Brewer and D.O. Edwards, J. Low Temp. Phys. **43**, 327 (1981).

8. C.J. Gorter and J.H. Mellink, Physica **15**, 285 (1949).

Figure Captions

1. Flow patterns associated with arbitrary vortex filaments in He II. The quantized-vortex filaments interact with themselves, with each other, and with boundaries through the circulating velocity fields.
    Physics Today, Feb. 1987.

2. Typical example of a vortex tangle in a rough channel, generated by numerical simulation as described in Section III. The microturbulence is driven by a superfluid field moving from left to right. Only a section of the tangle is shown. Phys.Rev.B, *38*, 2398 (1988).

3. Triad of vectors characterizing the instantaneous local configuration of the curve $\vec{s}(\xi, t)$. Phys.Rev.B, *31*, 5782 (1985).

4. Line gain and loss mechanisms. In (a) $\vec{v}_n - \vec{v}_s - \beta\vec{s}' \times \vec{s}''$ is pointing out of the plane of the figure, in (b) it is pointing into the plane.
    Phys.Rev.B, *31*, 5782 (1985).

5. Multiplication of singularities through the reconnection process. Here two vortex lines reconnect to form five. Phys.Rev.B, *38*, 2398 (1988).

6. Case study of the development of a vortex tangle in a real channel. Here, α = 0.10, corresponding to a temperature of about 1.6 K. The tangle is being driven by a superflow into the front face of the channel section shown. Phys.Rev.B, *38*, 2398 (1988).

7. Predicted mutual friction force as a function of temperature and driving velocity. The dots are the measurements of Brewer and Edwards[7] in a 0.0366 cm glass capillary. Phys.Rev.B, *38*, 2398 (1988).

Figure 1

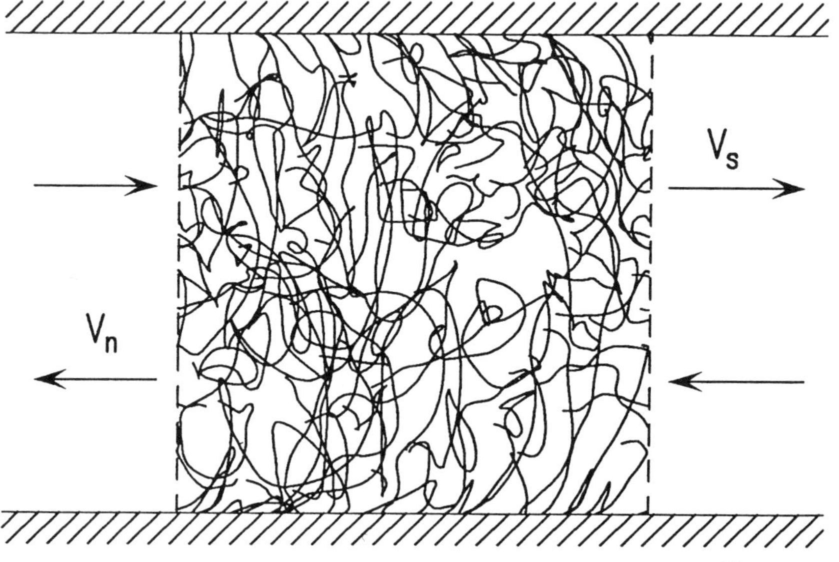

Figure 2

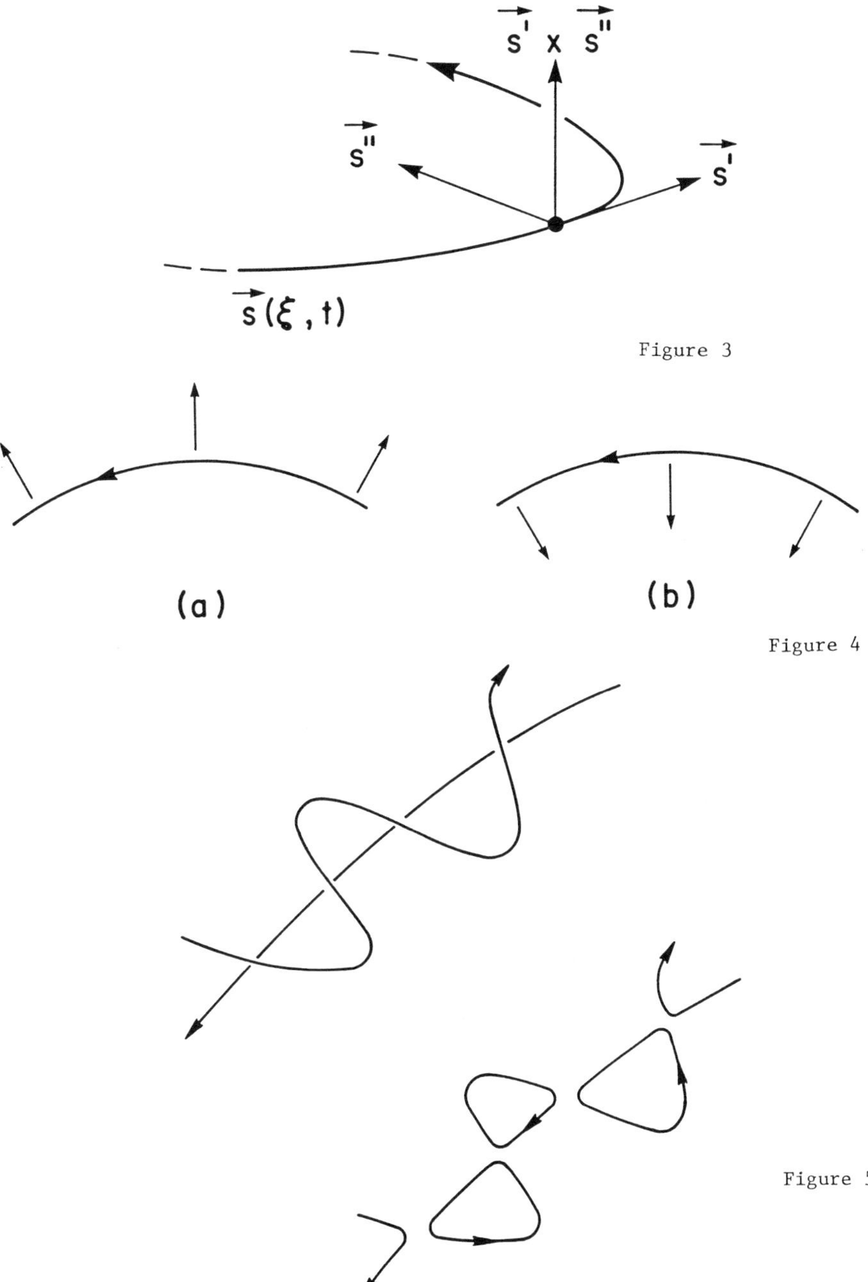

Figure 3

Figure 4

Figure 5

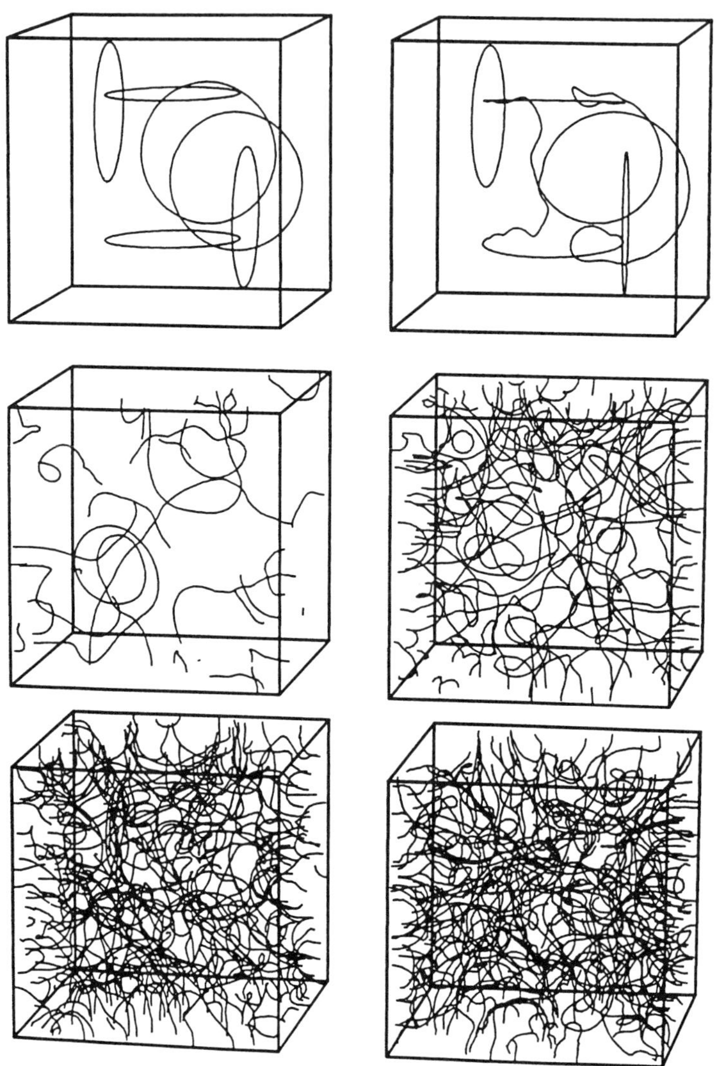

Figure 6

1.990
1.847
1.686
1.563
1.380
1.174
25
20
15
10
5
0
$\overline{F}_{sn}^{1/3}$ (gm cm^{-2}sec^{-2})$^{1/3}$
0   10   20   30   40   50
v_{ns} (cm sec^{-1})

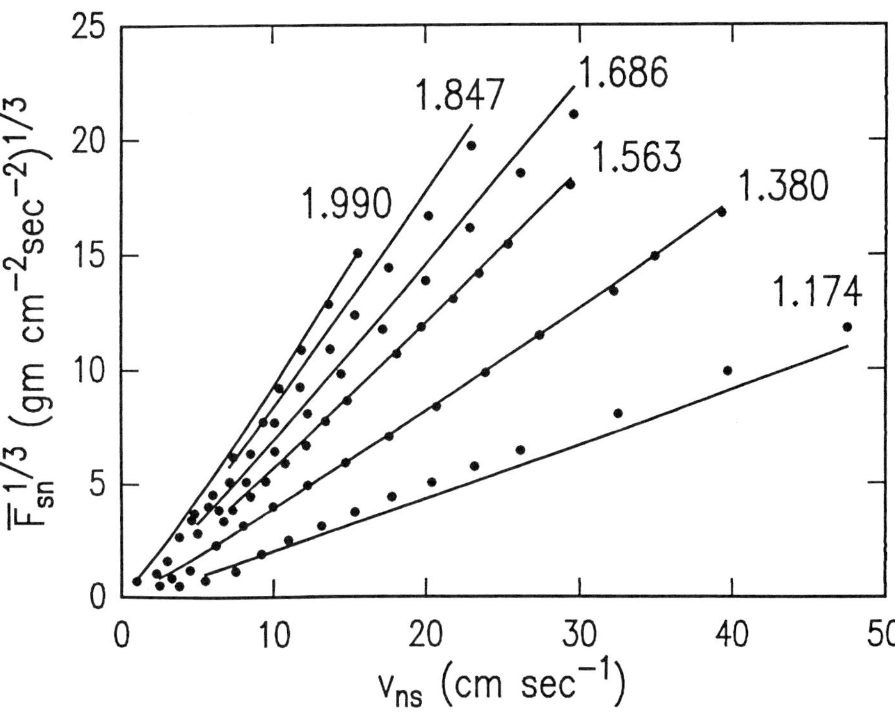

Figure 7

APPLICATION OF FLOW VISUALIZATION TECHNIQUE
TO SUPERFLOW EXPERIMENT

N. Ichikawa
Mechanical Engineering Laboratory, MITI, Tsukuba, Japan 305

M. Murakami
University of Tsukuba, Tsukuba, Japan 305

ABSTRACT

A flow visualization technique in He II was developed by the use of H_2-D_2 solid particles and hollow glass spheres as tracers. Thermal counter flow jet and large scale vortex rings in He II were investigated by flow visualization.

INTRODUCTION

Flow visualization technique has been developed in order to recognize flow patterns in various kinds of ordinary fluids. Qualitative feature of flow can be readily grasped by using the technique. In addition, quantitative understanding can also be reached from minute flow pattern examination with the development of image processing technique. It has been considered that flow visualization technique would be of use for understanding of flow phenomena in He II. Though some researchers [1-4] had attempted to visualize flows in He II, there had been few clear photographs of flow patterns.

There are some difficulties in applying conventional flow visualization techniques to He II. Solid particles must be used for visualization tracers because of cryogenic temperature. In addition, both density and viscosity of He II are very small. Therefore, it is very difficult to produce neutrally buoyant solid tracers in He II. This is the crucial point for successful flow visualization in He II.

In this study, the tracer method using hydrogen-deuterium solid particles and hollow glass spheres is applied to He II flow visualization. Details of production of tracer particles and some experimental results are described.

PRODUCTION OF TRACER PARTICLES

Hydrogen-Deuterium (H_2-D_2) Solid Particles

The application was first mentioned by Chopra and Brown in 1957 [1]. Most researchers who had tried flow visualization in He II might have used these particles as tracers. A schematic illustration of producing H_2-D_2 tracers is given in Fig.1. Premixed H_2 and D_2 gas is injected to He II free surface through heated injection tube. The exit of the tube is covered with a fine mesh screen to make the exit flow turbulent. The gas is solidified when it interacts with cold helium vapor and contacts He II at free surface. The size of solid particles at this moment is mostly the order of 1 μm. Another large wire net is set between the test space and the free surface in order to ensure uniform distribution of tracer particles. Then, particles may grow as a result of cohesion to the order of 0.1 mm being suitable for visualization. The net also serves to trap larger flakes and excess particles. After considerable trial and error, it is found that there are some empirical rules to produce satisfactory visualization condition. The first is the concentration ratio of H_2 to D_2. The composition of H_2-D_2 mixture is about 10 to 1 by the partial pressure. This is not the theoretically optimum mixing ratio,

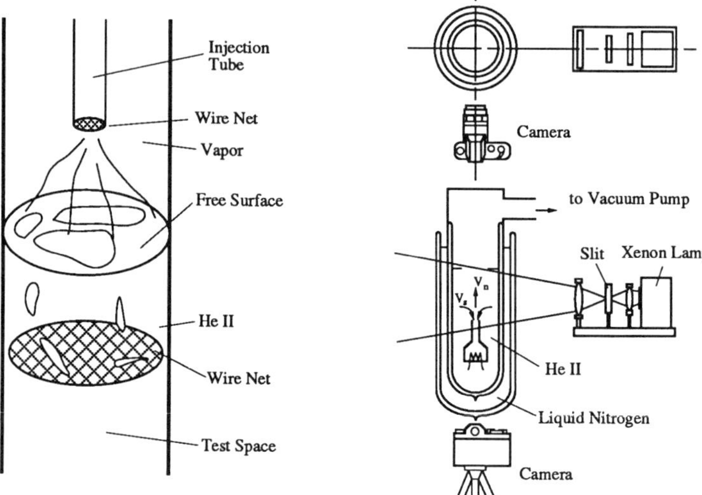

Fig.1   Schematics of producing of H_2-D_2 tracer particles.

Fig.2   Experimental setup.

which is about 1 to 1.   It is considered that the composition of solid particles may be different from that of gaseous mixture because of imperfect solidification of the two components with different solidification tendencies.   The second relates to the temperature and the relative position of the injection tube with respect to the free surface.   The temperature of the tip portion of the tube should be warmed up to about 80 K by a heater prior to an injection.   The distance between the tube exit and the He II free surface is about 3 cm.   It is necessary to renew particles every about 30 minutes because of rapid disappearance of tracer particles in the test space.   The gas quantity is the order of 1 μmol. per injection per 10 liters of He II.   These particles are used as tracers suspend in the test space in neutrally buoyant state.

Hollow glass spheres

It is commercially available low density composite material with an average density of 200 kg/m^3.   The diameters range from 20 to 100 μm.   They are very fine and light, thus the particle trap must be set in the vapor evacuation line to filter out particles from the vacuum pump.   Unfortunately, this sphere could not suspend in He II stably.   Therefore these spheres are used for different purpose from H_2-D_2 particles.

FLOW VISUALIZATION EXPERIMENT

Experimental Setup

Figure 2 is a schematic illustration of the experimental setup.   A 300 W Xenon lamp is employed as a light source.   A light sheet illuminates the plane including the jet axis.   The width of the light sheet is about 1 mm at the test space.

Visualization of Thermal Counter Flow Jet

The principle of thermal counter flow jet is shown in Fig.3. A heater is located at the bottom of the jet chamber. When heat is applied, the super component moves to the heater and the normal one flows out. Then a jet flow is formed of the normal component in the external field. Since solid particles follow the normal component, only normalfluid flow field can be observed. The inner diameter of the nozzle is 4.0 mm and the length of the nozzle is about 20 mm.

Prior to taking photographs appropriate trace particles are produced in He II when H_2-D_2 particles are used as tracers. Hollow glass spheres are sometimes used to observe the evolution of internal jet structure, which are put into the chamber before the experiments. Heat is applied stepwise to He II in the chamber. Photographs are taken after a steady state is reached.

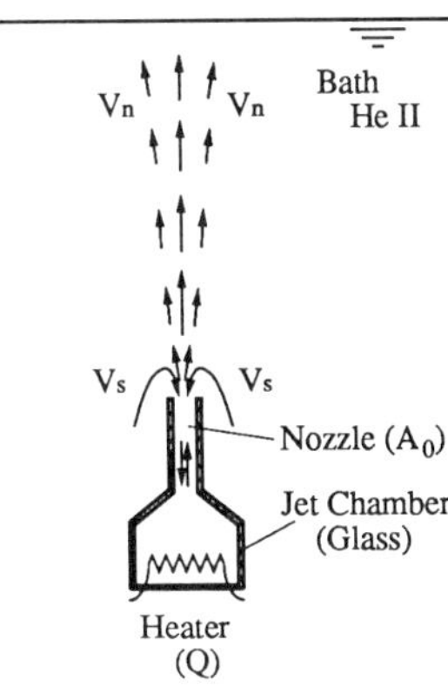

Fig.3   Principle of thermal counter flow jet in He II.

Examples of the thermal counter flow jet are shown in Fig.4 (a), (b) and (c). In the case of Fig.4 (a), H_2-D_2 solid particles are used as tracers. Heat input is small and jet is found to be laminar. The picture shown in Fig.4 (b) is the result when heat input

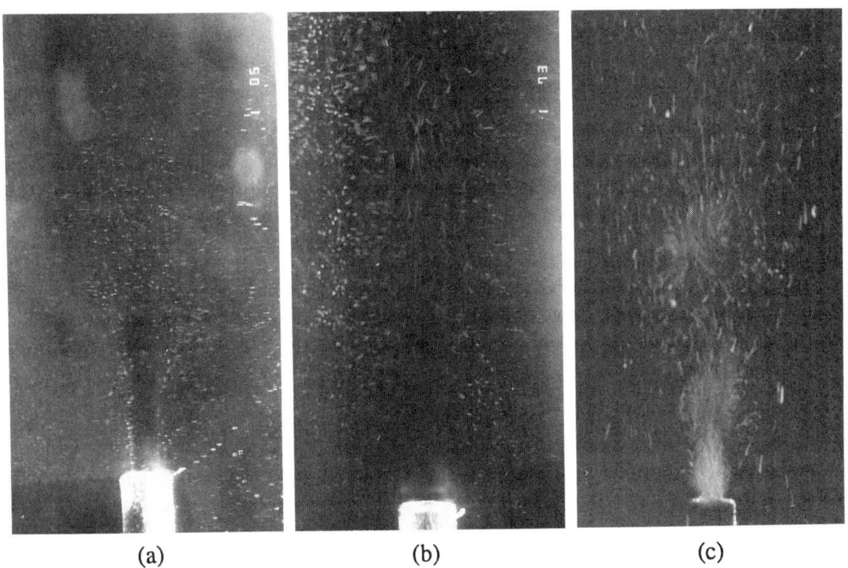

| (a) | (b) | (c) |

Fig.4   Pictures of thermal counter flow jet. *Cryogenics*, 29, Apr 1989, pp.438-443.
    (a) q=1.06x10^3 W/m^2, T_{bath}=1.99 K, H_2-D_2 particles, F2.8 1/2 s seventh fold exposure.
    (b) q=1.77x10^4 W/m^2, T_{bath}=1.95 K, H_2-D_2 particles, F2.8 1/2 s ninth fold exposure.
    (c) q=3.78x10^4 W/m^2, T_{bath}=1.90 K, Hollow glass spheres, F2.8 1/15 s exposure

212   N. Ichikawa and M. Murakami

is about 10 times greater than that in Fig.4 (a). A fully developed turbulent jet is seen accompanying large scale entrainment from the surroundings to the jet. Figure 4 (c) shows a result in the case of hollow glass spheres as tracers. Heat input is twice as large as the case shown in Fig.4 (b). Intermittent violent vortical motion superposed on a base turbulent jet is seen.

Particle trackability can be checked from these photographs. Figure 5 shows the comparison of the axial component of the average normal velocity between the theoretical prediction and those obtained from particle streaks near the nozzle exit in the photographs. The experimental values seem to agree fairly with the theoretical ones for velocities smaller than about 5 cm/sec.. The reason why the experimental data for larger velocities are below the theoretical line is considered to be actual reduction in flow velocity caused by superfluid turbulent behavior. The same feature is also observed in Laser Doppler Velocimeter measurements with smaller tracers [5].

Spatial velocity distributions can even be obtained from particle streaks measured from the photographs. Figure 6 shows the results for the fully turbulent jet as shown in Fig.4 (b). The abscissa expresses the dimensionless radius, r/b, and the ordinate does dimensionless axial velocity, V_n/V_{nmax}. The solid line represents Görtler's profile for viscous turbulent jets given by

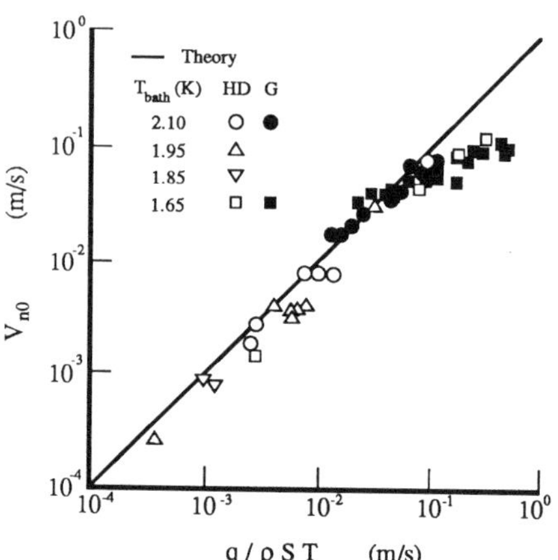

Fig.5   Particle trackability in He II.
Cryogenics, 29, Apr 1989, pp.438-443.

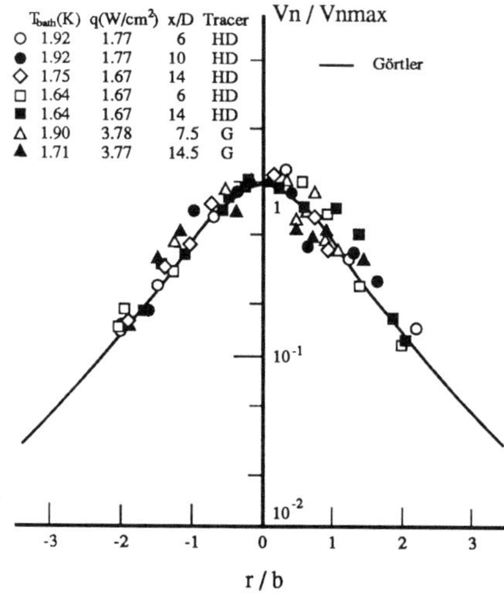

Fig.6   Velocity distribution in thermal counter flow jet.
HD, H_2-D_2 particles; G, Hollow glass spheres.
Cryogenics, 29, Apr 1989, pp.438-443.

$$V_n / V_{nmax} = 1 / ((1 + (\sqrt{2} - 1) \times (r/b)^2)^2), \qquad (1)$$

where V_n is the axial velocity of the normal component, V_{nmax} the velocity on the jet axis, r the radius and b the radius where $V_n = V_{nmax}/2$. This is based on the eddy viscosity assumption [6]. The agreement seems fair within the experimental error. It is considered that the velocity distribution of the normal component in the thermal counter flow jet is close to that of the ordinary turbulent jet.

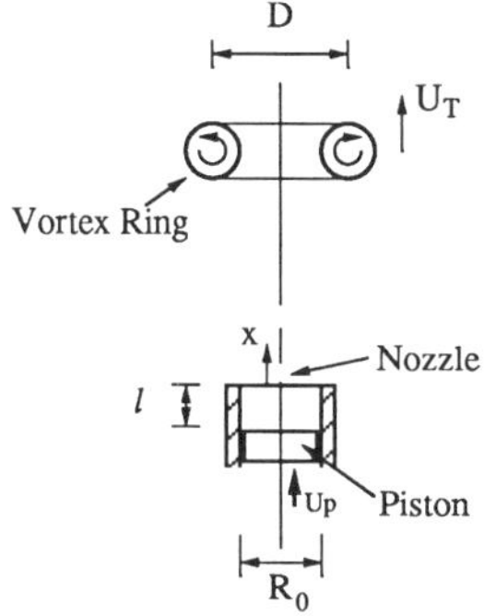

Fig.7   Schematics of large scale vortex ring translation in He II. *Proc 12th ICEC*, 1988.

Visualization of Large Scale Vortex Ring

Figure 7 illustrates a vortex generator and a vortex ring. The vortex ring is generated by pushing a piston in the vortex generator. The piston is driven by a linear drive mechanism mounted outside the Dewar. The setup for visualization is the same as that of the thermal counter flow jet. H_2-D_2 solid particles are used as tracers.

Translational motion of a large scale vortex ring in He II is seen clearly in Fig.8 (a) and (b). From the detail analysis of these photographs, it is found that the vortex ring in He II translate quite similarly to turbulent ones in viscous fluids [7].

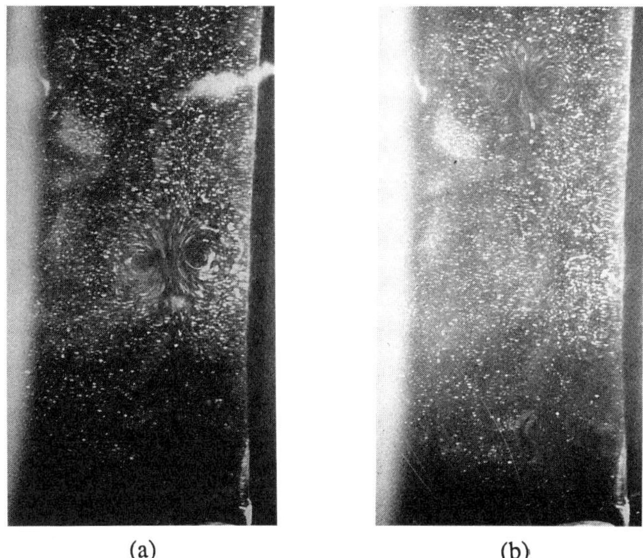

(a)   (b)

Fig.8   Pictures of translation of vortex ring.
T_{bath}=2.15 K, H_2-D_2 particles, F2.8 1/8 s exposure.

CONCLUSIONS

Application of flow visualization technique in He II has been successfully developed by the use of H_2-D_2 solid particles and hollow glass spheres as tracers. Flows of thermal counter flow jet and large scale vortex ring translation are clearly visualized. Particle trackability of both kinds of tracers is fairly good. Velocity distribution of normal component in thermal counter flow jet is found to be similar to that of ordinary turbulent jet.

REFERENCES

1. K. L. Chopra and J. B. Brown, Phys. Rev. 108, 157 (1957).
2. T. A. Kitchens et al., Phys. Rev. Lett. 14, 942 (1965).
3. D. Y. Chung and P. R. Critchlow, Phys. Rev. Lett. 14, 892 (1965).
4. H. W. Liepmann and G. A. Laguna, Ann. Rev. Fluid Mech. 16, 139 (1984).
5. T. Yamazaki and M. Murakami, Cryo. Eng. 23, 49 (1988) (in Japanese).
6. N. Rajaratnam, Turbulent Jets, Elsevier Sci., (1976).
7. M. Murakami and M. Hanada, Proc. of 12th ICEC, (1988).

LASER DOPPLER VELOCIMETER APPLIED TO SUPERFLOW MEASUREMENT

M. MURAKAMI, T. YAMAZAKI, A. NAKANO and H. NAKAI
Inst. of Eng. Mechanics, University of Tsukuba, Tsukuba 305 Japan

ABSTRACT

The application of a Laser Doppler velocimeter (LDV) to He II
flow measurement was attempted for sufficiently high accuracy and
resolution in both time and space. The detail of the application
technique and some results measured with the LDV for the thermal
counterflow jet are described here to demonstrate the applicability
of an LDV as major diagnostic means for the He II flow facility.

INTRODUCTION

In order to investigate He II flow phenomena from the fluid-
mechanical point of view, it is absolutely necessary to measure the
flow velocity as well as such state variables as the temperature and
the pressure. However, only primitive techniques are available for
the measurement, by which the mean velocity reduced from the volume
flow rate can, at best, be obtained. The application of an LDV may
make a breakthrough to the field. The attempt had led to a success
after a considerable try and error in suitable tracer particle
production, and then the LDV has been applied to the measurement of
the thermal counterflow jet in He II as an example. The thermal
counterflow is one of the overwhelming dominant modes of heat
transfer in He II. The flow is suitable for investigation of the
intrinsic nature of He II flows because of being free from effects of
solid boundaries, and has been investigated by many researchers[1,2,3],
with whose data the LDV data can be compared[4].

In this report, some detail of the LDV application is described
to discuss potential problems in the application to the He II flow
facility. And the result of the thermal counterflow jet experiment
is presented, which indicates that the LDV system with hydrogen-
deuterium solid particles as tracers has, in fact, high time-
resolution and that the particles well follow the normalfluid
component[5]. Regular oscillation in the normal fluid velocity with
frequencies of 10 to 30 Hz is detected, which is caused as a result
of the 2nd sound Helmholtz oscillation in the jet chamber system.

LDV APPLICATION

LDV System
The LDV system (Fig. 1) works in the double beam forward
scattering mode with a 15 mW He-Ne laser. Two beams from a beam
splitter produce Young's interference fringes in an intersection
volume, that is the measuring volume. When a solid particle passes
through the volume, it scatters light which undergoes a modulation
with a frequency determined by the particle velocity divided by the
fringe interval. Each event produces a burst signal, which is
detected by a photomultiplier and then is processed with either a
tracker or a counter to convert the frequency of a burst signal to

the velocity. The data from processors can be regarded as the normalfluid velocity due to the viscous nature, though velocities of tracer particles are, in fact, measured with an LDV. Accordingly, tracers are needed for the LDV application. They should be prepared with special care so as to be neutrally buoyant in He II, and to have satisfactory trackability. Hydrogen–deuterium solid particles as used in the flow visualization but much smaller in sizes, say the order of $1\,\mu$m, are employed as tracer particles. A Bragg cell type frequency shifter is utilized to determine the polarity of the flow velocity and to enhance the S/N ratio by minimizing the effect of multi-scattered stray light from surfaces of the window glasses and the cryostat wall, that is called the pedestal.

Cryostat for LDV Application

The cryostat with two windows each of which consists of three sheets of optical glass was designed in order to apply an LDV as seen in Fig. 2. The innermost two are superfluid tight and the intermediate two are coated with IR absorbing thin film and cooled to the LN_2 temperature to minimize 300 K thermal radiation heat input from the outside. The cryostat is equipped with a traversing mechanism for adjusting the vertical position of the jet nozzle. The inner diameter of the measuring section is 16 cm.

THERMAL COUNTERFLOW JET

A jet composed of the normal component emerges from a nozzle as a result of the thermal counterflow, in which the normal component flows from the heat source towards a heat sink, this is the ambient He II, and the super does in the opposite direction under the constraint that the net mass flow rate is zero,

$$\rho_s U_s + \rho_n U_n = 0 , \tag{1}$$

where U_s and U_n are the velocities of the super and normal fluids, and ρ_s and ρ_n are the super and normal densities, respectively, as schematically illustrated in Fig. 3. Since heat is transferred only by the normal fluid, the average velocity is evaluated at the nozzle exit as follows;

$$U_{n0} = (Q/A_0) / \rho ST = q / \rho ST. \tag{2}$$

Here, Q is the rate of heat generation, A_0 is the cross sectional area of the nozzle, S and T the specific entropy and the temperature of He II, and q the heat flux, Q/A_0. The normal flow velocity is measured with the LDV. The jet chamber made of glass contains a heater and has a straight circular nozzle with a constant diameter. In the present investigation, two types of chambers are utilized, of which description is given in Fig. 4.

EXPERIMENTAL RESULT AND DISCUSSION

Trackability Test

Prior to the thermal counterflow jet experiment, the

trackability of the H_2-D_2 particles to the normal fluid velocity was tested by using a belows pump flow system which was capable of generating known mean velocities in both liquid nitrogen and He II. The velocity of jet is measured with the LDV near the exit of the nozzle. The result is shown in Fig. 5, where the measured data are plotted against calibrated values. The agreement is satisfactory up to 15 cm/s which is the maximum of the pump performance. This result indicates that H_2-D_2 particles possess sufficient trackability to the normal fluid flow even in He II.

Temperature Measurement
     The temperature difference between the inside and outside of the jet chamber is plotted against heat flux, q, in Fig. 6. Two flow states are clearly noted according to the magnitude of q. $\triangle T$ increases roughly in proportion to q for small q, while it strongly does proportionally to q^3 in case of q above a critical value, q_c, about 0.5 W/cm^2. It is obvious that the cubic dependence results from the superfluid breakdown. These results suggest that the flows in the nozzle are in the superfluid turbulent state in cases $q > q_c$.

Average Normal Fluid Velocity
     A comparison is made between a simple theoretical prediction of the axial component of the average normal fluid velocity given by Eq. (2) and the experimental results, shown in Fig. 7. It is seen that the agreement is satisfactory for smaller velocities corresponding to the subcritical heating cases in accordance with the temperature measurement result. The discrepancy between them becomes noticeable above 10 cm/s. This can be attributed not to poor trackability of particles to the normal flow but to degradation of flow caused by additional dissipation mechanism such as turbulent motions. The trackability test shows no such discrepancy up to 15 cm/s, that is the upper limit of the pump performance [5]. It also seems that the phenomena cannot be characterized by a single parameter, $q/\rho ST$, but rather depend separately on T and q.

Transient Velocity Variation
     Typical transient velocity variations measured with the LDV are given in Fig. 8 a and b for two cases, $q < q_c$ and $q \gtrsim q_c$, respectively, where the time variation of the axial component of the normal velocity measured at the nozzle exit is plotted against time elapsed from the heater switch-on. In the former case, the mean velocity grows gradually. Very interesting point is appearance of regular velocity oscillation observed in the initial stage, which will be discussed later. The Reynolds numbers defined by

$$Re = \rho d_0 U_{n0} / \eta_n \quad (3), \qquad Re_n = \rho_n d_0 U_{n0} / \eta_n, \qquad (4)$$

are 1500 and 480, respectively. Here, d_0 is the nozzle diameter, and η_n the normal viscosity. The result suggests the normal fluid flow is laminar in the nozzle. Fig. 8 b is another velocity record in the case of moderately large heat input. Far larger velocity and much rapid response to the heater excitation as compared with the subcritical case, turbulent behavior and no regular oscillation are noticeable under the supercritical heating condition. The flow may be in the superfluid turbulent situation. Average velocity, about 6 cm/s, is still considerably smaller than the simple prediction.

<u>Second Sound Helmholtz Oscillation</u> [4]

Regular oscillation in the velocity is recorded in the initial stage in the case of small q as seen in Fig. 8 a, which is found to be the second sound Helmholtz oscillation. The fundamental frequency is plotted against the bath temperature for both jet chambers in Fig. 9. The solid and the dashed lines are the theoretical prediction f_0 given by

$$f_0 = (a_2/2\pi)(A_0/L_0 V_c)^{1/2} \tag{5}$$

where

$$a_2 = (\rho_s S^2 T/\rho_n C)^{1/2} .$$

Here a_2 is the speed of the second sound and C is the specific heat. A_0 and L_0 are the cross sectional area and the length of the nozzle and V_c is the volume of the jet chamber. It is of importance to note that the frequency is given as a function of temperature and geometry of the chamber independently of q. This is a standing second sound wave in the chamber, and thus the velocity oscillation can hardly detected in the downstream region outside the chamber. The Helmholtz oscillation is sometimes observed after the heater power off, though only turbulent-like jet appears during heating.

SUMMARY

High resolution measurement of the normal fluid flow velocity has been successfully made with a laser Doppler velocimeter. Solid H_2-D_2 particles are found to be suitable choice for neutrally buoyant tracers in He II. On the other hand, several possible problem area is pointed out for the LDV application to the large scale liquid helium flow facilities.
i)   Suitable production and maintenance of tracer particles.
ii) Optical windows may be tremendous heat source to the He II temperature environment in a cryostat and may cause leak problem. Application of fibre optics instead of windows may partly solve these problems.
iii) Alignment of LDV optics may be hard in large facility and cold test section.
iv) Surface contamination with tracer particles on a model and windows.

REFERENCES

1. G. A. Laguna, Phys. Rev. B12, 4874 (1975).
2. P. E. Dimotakis and G. A. Laguna, Phys. Rev. B15 5240 (1977).
3. M. Murakami and N. Ichikawa, Cryogenics 29-4, 438 (1989).
4. M. Murakami, T. Yamazaki and H. Nakai, Cryogenics 29-12 1143 (1989).

5. T. Yamazaki and M. Murakami, J. Cryo. Soc. Japan 23, 49 (1988) (in Japanese).

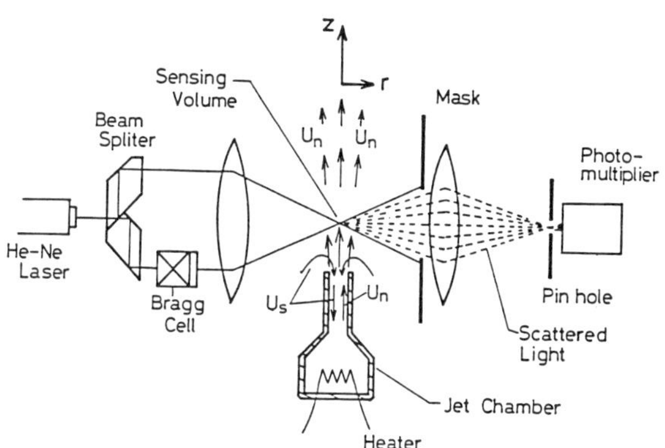

Fig. 1
LDV optical configuration

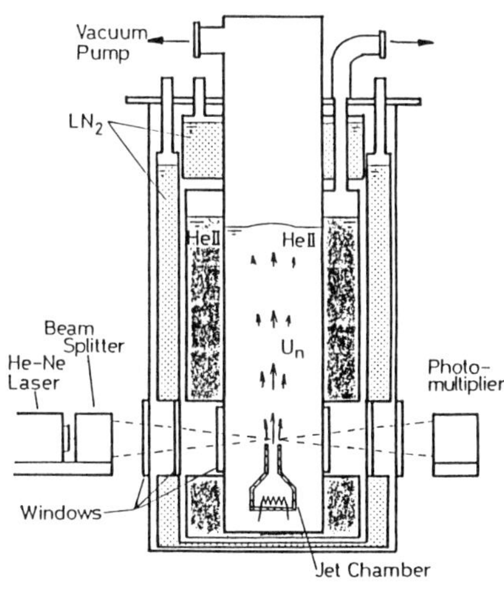

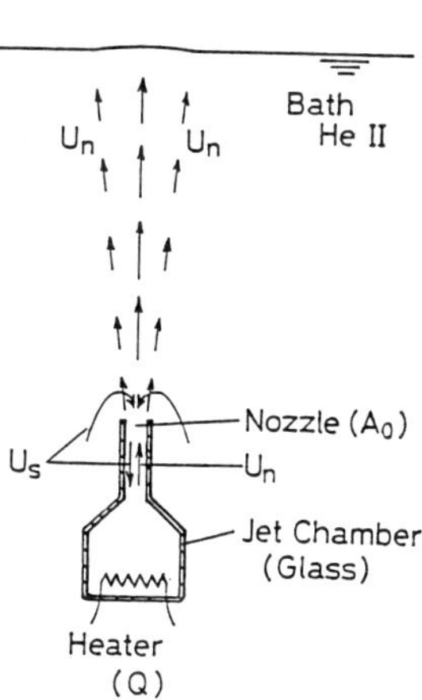

Fig. 2 Cryostat designed for optical measurement

Fig. 3 Schematic of thermal counterflow jet

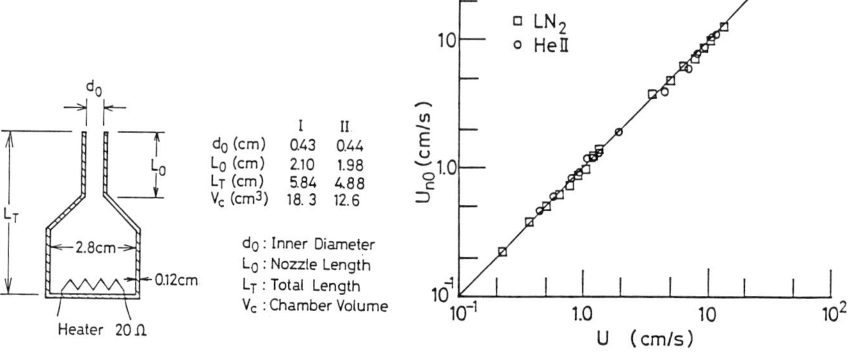

Fig. 4 Description of jet chambers

Fig. 5 Trackability test result. Comparison of LDV data with calibrated velocity.

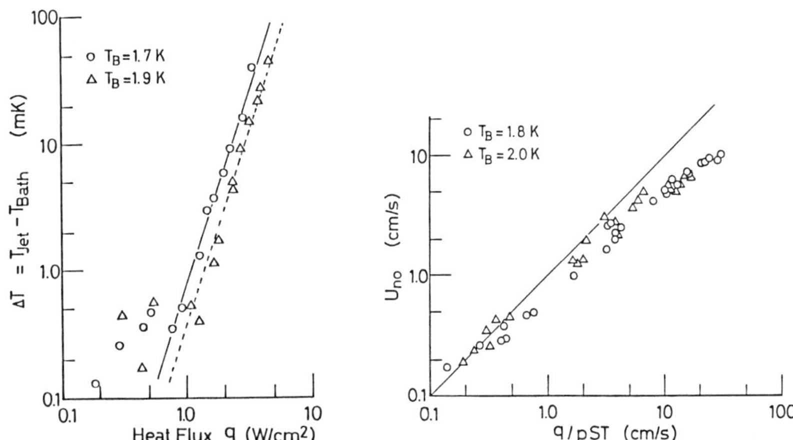

Fig. 6 Temperature difference as a function of q.

Fig. 7 Comparison of LDV result with simple theoretical prediction.

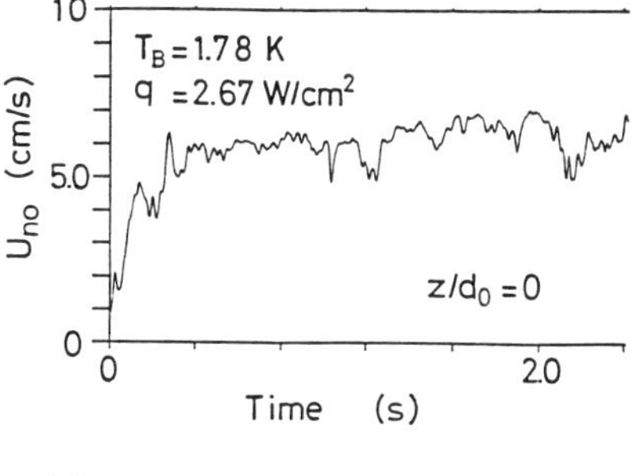

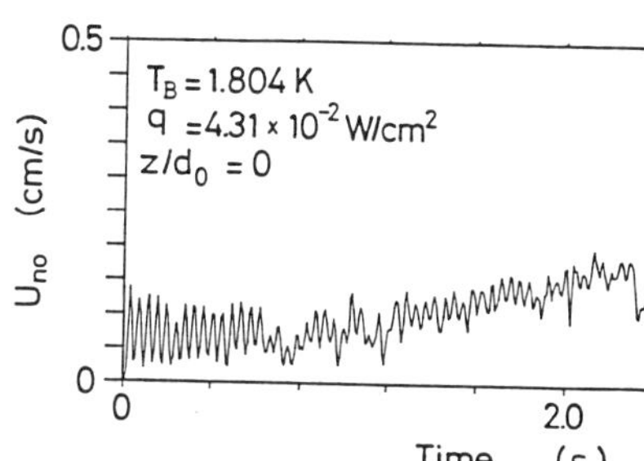

Fig. 8 Transient LDV record.
a:subcritical heating;
b:supercritical heating.

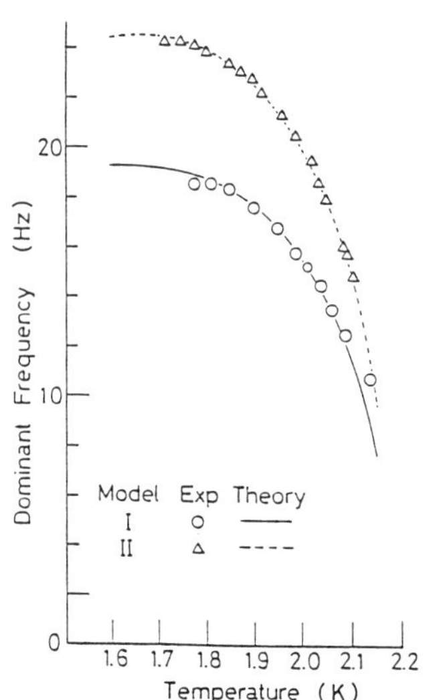

Fig. 9 Frequency of 2nd sound
Helmholtz oscillation as a
function of temperature.
Data are compared with
theoretical prediction.

EXPERIMENTAL INVESTIGATIONS OF HE II FLOWS AT HIGH REYNOLDS NUMBER

S. W. Van Sciver
Applied Superconductivity Center
University of Wisconsin
Madison, WI 53706

ABSTRACT

Fluid dynamics studies of He II at high Reynolds number ($Re_d > 10^6$) reveal characteristics which are best interpreted in terms of classical scaling relationships. In particular, the smooth tube friction factor is seen to correlate with the Von Karman-Nikuradse formulation. Also, the performance of a centrifugal pump is unchanged whether being used with He I or He II. These effects are expected to result provided the He II possesses a viscous sublayer and that the drag is determined by laminar flow within this layer. On the other hand, heat transfer in He II is substantially different from that of He I because of the unique internal convection mechanism present in this quantum fluid. These experiments are performed in the University of Wisconsin liquid helium flow facility which has unique capabilities of He II temperature, pressure and flow.

INTRODUCTION

Liquid helium is the required coolant for numerous large scale technical devices. Forced flow He II (superfluid helium) is being considered for a subset of these devices: in particular several large scale superconducting magnet systems and space based infrared telescope facilities. The need to understand the fluid mechanics of forced flow He II for the design of these systems has inspired recent experimentation at high Reynolds number ($Re_d > 10^6$).

Forced flow He II is also under consideration as an operating fluid for a liquid helium wind tunnel[1], which is the subject of the present workshop. The operation of such a wind tunnel and interpretation of experimental data will require a thorough understanding of the behavior of He II at extremely high Reynolds number. The present report describes current understanding of high Re_d He II and facilities at the University of Wisconsin utilized in its study.

LIQUID HELIUM FLOW FACILITY (LHFF)

The primary instrument for the study of forced flow He II is the liquid helium flow facility. The LHFF is designed to allow a variety of experiments which test important helium fluid dynamics behavior on a scale close to that involved in large scale applications. To achieve this goal, we chose a horizontal dewar configuration with a cold bore access on either end. A nominal length of five meters was selected with a sufficiently large inner diameter to allow insertion of various tubing configurations, flow metering devices and heat exchangers. Further, to minimize consumption of liquid helium, the dewar design includes two actively cooled shields; one cooled by LN$_2$ to 77 K and one at 4.5 K maintained by a closed cycle helium refrigerator. Helium flow is to be provided by a cold centrifugal pump.

The schematic representation of the LHFF, as it is currently configured with the liquid helium pump and one vertical stack insert including heat exchanger, is shown in Fig. 1. An alternative mode of operation can be achieved by replacing the pump with a second vertical stack. Also, in the current configuration, the stack insert can be exchanged so that other heat exchangers or helium reservoirs can be inserted into the system.

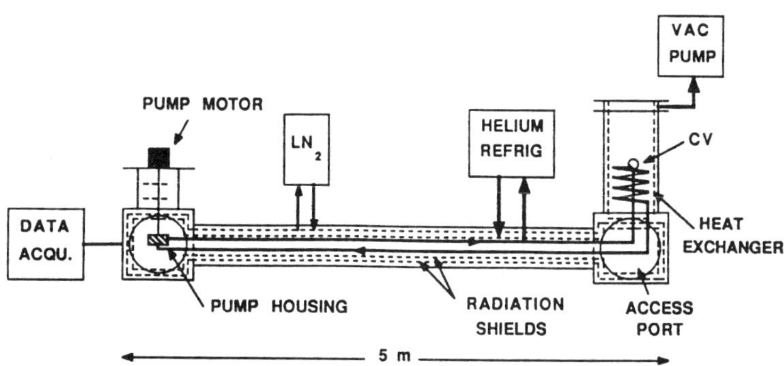

Fig. 1. Schematic of liquid helium flow facility (LHFF).

The actual dimensions of the horizontal dewar assembly are 5 m length and 20 cm ID. Each end box has three access ports which are 22.9 cm in diameter. These six 4.5 K flanges have indium seals to separate the isolation vacuum from the experimental vacuum space. The inner shield is cooled by a forced flow loop connected to a CTI-1400 refrigerator operating in closed cycle. The liquid nitrogen shield is a 30 liter storage reservoir located on the top of the dewar.

The liquid helium pump, located at one end of the dewar assembly, provides the forced flow for experimental investigations. It consists of a 6.86 cm diameter impellar with six straight blades housed in a volute casing. The pump is supplied with four different impellar housings dependent on the desired operating pressure and flow rate. A schematic representation of the pump housing is shown in Fig. 2. The impellar is connected to a room temperature motor over a 44.5 cm, 2.5 cm OD tubular shaft. The shaft is supported by gas lubricated ball bearings and rotates at moderate frequencies (f ~ 100 Hz). The pump is designed to supply a liquid helium flow of up to 100 gm/s with a thermodynamic efficiency of 60%. It pumps single and two-phase normal helium as well as He II.[2]

The vertical stack end of the LHFF is designed to allow insertion of different liquid helium reservoirs including heat exchanger assemblies. At present three such reservoirs exist, one open bath for experiments near saturation and two others with heat exchangers, one designed for operation at 4.2 K and the other with He II. A schematic representation of the He II

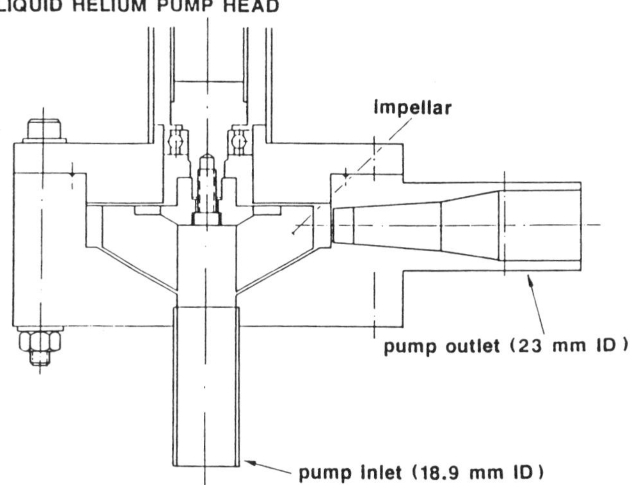

Fig. 2. Housing and impellar of the liquid helium pump.

heat exchanger assembly is shown in Fig. 3.

The He II heat exchanger system consists of two separate heat exchangers. The liquid-vapor heat exchanger sits in the upper reservoir and is used to precool the liquid which is injected into the saturated side of the liquid-liquid heat exchanger. The liquid-liquid heat exchanger is of tube-in-shell design.[3] Connections to the flow loop of interest are made using the two indium O-ring flanges located at either end of the heat exchanger section. The heat exchanger is equipped with a high conductance back pressure control valve (CV) to allow variable flow. Also, the system has temperature sensors at the inlet and outlet to monitor the heat exchanger performance. As designed, this heat exchanger system is capable of removing 10 watts at 1.8 K.

The procedure to be followed in preparing to run the LHFF begins by installing the test section within the inner dewar. To date, investigations have been limited to tubular test sections with varying geometrical cross section. All connections between the pump, heat exchanger and loop are made using indium O-ring flanges. Vacuum is then established in the dewar jacket after which cooldown commences with the experimental section surrounded in low pressure helium gas. Initial cooldown is accomplished by filling the LN$_2$ reservoir. After approximately two days the helium refrigerator begins cooling the inner shield. By this method, the heat leak to the 4.5 K temperature shield is minimized.   With the LN$_2$ shield being cooled, there is

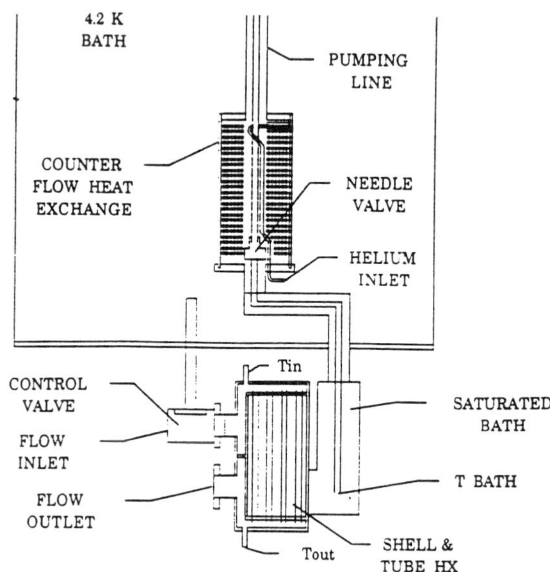

Fig. 3. He II tube-in-shell heat exchanger assembly.

only a gradual temperature decrease of the inner shield and pump components. However, after the refrigerator is turned on, the pump and inner shield cool rapidly. The system is ready for transfer of liquid helium after an additional two days of cooling.

A unique characteristic of the LHFF is that the helium circulation pump is not located in a reservoir of liquid. This design feature requires a certain procedure during cooldown. Most important is that the pump must be primed by transferring liquid directly through the circulation loop. Once primed, the pump is able to circulate the helium but because of residual heat leaks to the system, the helium tends to be two phase. This difficulty can be alleviated by beginning circulation in the He II regime since the heat conductivity of the static helium within the loop is sufficient to condense any residual vapor. The loop can then be isolated from the saturated reservoir and pressurized. If experiments in single phase normal helium are to be performed, the loop can be easily warmed above T_λ by circulating the liquid through the pump with the heat exchanger regulated in the desired temperature range.

Estimates of the heat leaks to 77 and 4.5 K are made by different methods. The 77 K heat leak is determined by simply measuring the nitrogen gas evolved from the storage reservoir. In steady state with the facility fully cold and operational, we have determined this heat leak by about 27 watts. To estimate the 4.5 K heat leak, we calculate the total cold mass and measure the time rate of change of the shield temperature with the active cooling removed. This dynamic heat leak is estimated to be six watts. We believe that the rather large heat leak at low temperatures is a combination of the helium pump having room temperature components connected through a rather short shaft to the low temperature region and that the facility has a large number of electronic leads, all of which are heat sunk at 4.5 K.

EXPERIMENTS

As designed the LHFF is capable of a variety of low temperature helium fluid dynamics studies. In operation with He II, it has been used in measurements of 1) pressure drop through smooth tubes and a variety of flow elements, 2) heat transfer to forced flow He II including heat exchanger performance, and 3) the operation of liquid pumps for circulation of He II. The results of these findings are discussed in more detail below:

Pressure drop studies

Measurements of pressure drop in smooth tubes of various diameters between 3 and 8 mm have yielded data which is compared to the classical Fanning friction factor expression

$$f_s = \frac{\pi^2}{32} \frac{D^5 \rho}{\dot{m}^2} \frac{\Delta P}{L} \tag{1}$$

where
    f_s - Fanning friction factor
    D - tube diameter
    L - tube length
    $\dot{m}$ - mass flow rate
    ΔP - total pressure drop
    ρ - fluid density

The Reynolds number has been calculated using the traditional definition,

$$Re_D = \frac{4\dot{m}}{\pi D \eta_n} \tag{2}$$

except that the relevant viscosity η_n is that of the normal fluid component. These results have been compared to previous measurements of pressure drop in coiled tubing. In the case of curved tubing there exists an empirical expression to account for curvature[4]

$$f_c = f_s \left[\left(\frac{D}{2R} \right)^2 Re_D \right]^{1/20} \tag{3}$$

where R is the radius of curvature of the coiled section and f_c and f_s are friction factors for the curved and straight sections respectively.

The straight and corrected curved tube results for 1.8 K He II are displayed in Fig. 4. Also shown in the figure is the Von Karman-Nikuradse smooth tube correlation

$$\frac{1}{f^{1/2}} = 1.737\, ln\left(Re_D\; f^{1/2}\right) - 0.396 \tag{4}$$

which is used for classical fluids at $Re_D > 10^5$. The agreement among the various sets of data is reasonably good, although the curved tube data is systematically slightly below those obtained for the straight section. It is possible that the correlation, eq. (3), overcorrects for tubing curvature.

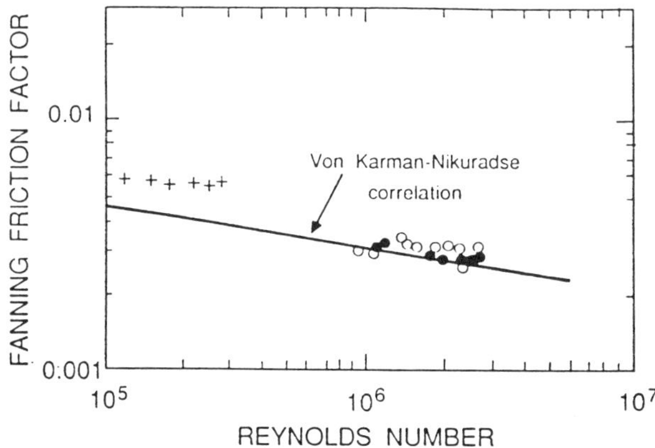

Fig. 4. Friction factor data for He II flow.

The observed agreement between the classical fluid pressure drop correlation and the behavior of He II may seem surprising when considering the different physical pictures used to describe the turbulent state in these fluids. Turbulent classical fluids contain macroscopic eddys which flatten the velocity distribution in pipe flow causing the viscous sublayer to thin. The velocity distribution in the laminar viscous sublayer near the wall controls the pressure drop. Using the language of the two fluid model for He II, the turbulent state is seen to consist of a tangled mass of quantized vortex lines for the superfluid component interacting with a turbulent normal fluid component. The interaction between these two fluids is responsible for the unique non-ideal heat transport behavior of He II and appears to be controlled by the relative velocities of the two fluid components, $(v_s - v_n)$. For adiabatic flow, the relative velocity between the two fluid components should be small compared to the overall fluid velocity; however, the interaction between the two fluid components appears strong enough to ensure that there is macroscopic turbulence of the fluid similar to that occurring in classical fluids. One can suggest that the apparent classical behavior of the friction factor in He II may result from the normal fluid component controlling the wall sheer stress in a viscous sublayer which is analogous to that observed in classical fluids. Further work will be required to fully evaluate this hypothesis.

HEAT TRANSFER STUDIES

Heat transfer to forced flow He II is important for the technical application of this unique fluid. Within the context of heat transfer, there are essentially two issues: 1) heat transport by the bulk fluid and 2) heat transfer between solids (usually metal) and the flowing He II. In the former case, the unique aspect is the existence of counterflow within the two fluids. On the other hand, solid-He II heat transfer is primarily determined by a process known as Kapitza conductance.

Heat transport in forced flow He II is most often analyzed in terms of the He II energy equation[6], which for one dimensional flow is written

$$\frac{\partial}{\partial x} \left[\frac{1}{f} \left(\frac{\partial T}{\partial x} \right)^{1/3} \right] - \rho v \frac{\partial h}{\partial x} = \rho \frac{\partial h}{\partial t} - q_{gen} \tag{5}$$

where h is the specific enthalpy and q_{gen} the applied heat flux. The quantity $f = A\rho_n/\rho_s^3 S^4 T^3$ is a temperature and pressure dependent effective thermal resistance for the He II. It is brought about by the mutual friction between the two fluid components. The time independent form to eq. (5) can be shown to lead to a dimensionless variable which for constant properties is written

$$K = \rho C v \ (fL)^{1/3} \ \Delta T^{2/3} \tag{6}$$

where in this case ΔT is the temperature difference along the channel. The dimensionless variable K can be thought to represent the ratio of heat carried by forced convection to that by internal convection. It is analogous to the Peclet number in normal fluids. The above formulation has been shown to be consistent with a number of heat transfer experiments in forced flow He II.

The second issue of heat transfer between metals and forced flow He II has only recently been investigated. In general experiments have shown that the heat transfer process is dominated by Kapitza conductance and is not significantly affected by fluid velocity. Heat transfer data at moderately high heat fluxes has been shown to best correlate with the expression

$$q = \alpha \ (T_s^n - T_b^n) \tag{7}$$

where the exponent n should theoretically be equal to four but experimentally is found to be somewhat less than that. The temperatures in eq. (7) represent that of the surface T_s and adjacent fluid, T_b. The coefficient α depends to a considerable degree on the characteristics of the material surface. Recent measurements on copper give values of $\alpha \cong$ 1000 W/m^2K^3 for n = 3, consistent with published results.[6]

An interesting and important application of the heat transfer processes with forced flow He II is in the development of heat exchangers. Proper design of these devices requires an understanding of both heat transfer and transport processes in He II. For example, the tube-in-shell heat exchanger used in the present set of experiments was designed using a semi-analytical calculation based on the He II energy equation with heat transfer.[7] More recently, a numerical analysis has been performed in conjunction with a series of experiments to identify the complete approach to design and development of tube-in-shell heat exchangers.[8] As a rule, heat exchanger performance in He II is degraded due to internal convection heat transport. Thus, conventional heat exchanger design criteria are inappropriate. The best approach, to minimize axial heat transport, is to make the heat exchanger as long as possible within the constraints of allowable pressure drop.

LOW TEMPERATURE PUMP CHARACTERISTICS

The centrifugal pump installed in the LHFF is of conventional design and does not contain any special features for use with He II. It is, therefore, of interest to determine whether such a pump will perform suitably when utilized with He II.

Throughout the series of experiments, we have found the centrifugal pump to perform similarly whether it is used with He I or He II. To be more quantitative we must consider the performance parameters associated with pump operation; the head coefficient, Ψ and the flow coefficient Φ are

$$\Psi = \frac{\Delta P}{\rho N^2 \pi^2 D^2} \tag{8}$$

and

$$\Phi = \frac{\dot{V}}{\pi D N A} \tag{9}$$

where ΔP is the pressure head
   N is the pump rotation frequency (Hz)
   D is the impellar diameter
   A is the diffuser area
   V is the volumetric flow rate

These non-dimensional terms permit the generation of a universal curve for the pump. In addition we define the efficiency of the pump,

$$\eta = \frac{\Delta P \dot{V}}{Q} \tag{10}$$

where Q is the total heat dissipated by the pump in moving the fluid. In calculating Q we have subtracted the steady state heat leak associated with the pump motor housing being at room temperature.

Figure 5 displays the measured head coefficient and efficiency for He

II at 1.8 K versus flow coefficient along with a performance prediction from the pump manufacturer. The performance characteristics are close to expected which adds further credence to the suggestion that the hydrodynamic character of turbulent He II is similar to that of classical fluids. We believe that this observation is consistent with the pressure drop characteristics of turbulent forced flow He II. The interaction between the fluid and impellar in a centrifugal pump is largely determined by frictional drag. As discussed above, the frictional interaction is primarily controlled by the flow of fluid in the viscous sublayer, which is essentially unchanged between He I and He II.

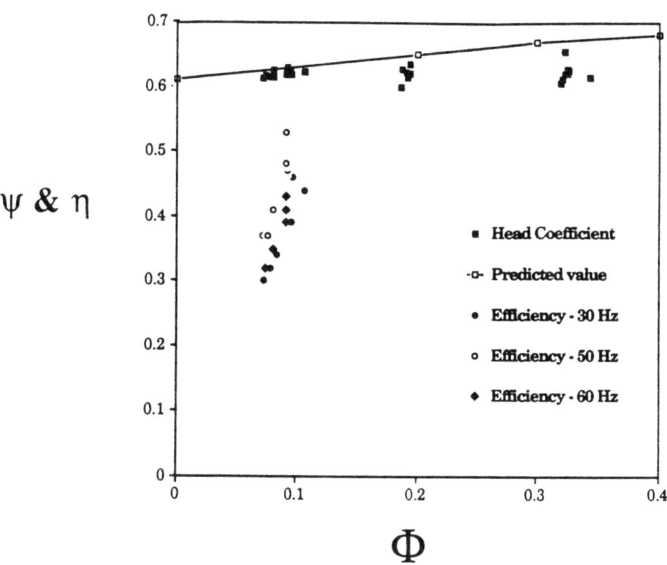

Fig. 5. Performance parameters for liquid helium pump.

CONCLUSIONS

Measurements of pressure drop in very turbulent He II have shown characteristics which can be best described by classical fluid correlations. Similarly, performance measurements on a centrifugal pump give values near those expected for classical fluids. This behavior is believed to result from the turbulent He II possessing a viscous sublayer, which determines the frictional interaction through the laminar normal fluid viscous drag. Should this hypothesis continue to be supported by experiment, it suggests that all drag measurements involving a viscous interaction will have essentially similar behavior in He II as in He I or other fluids. By contrast, the heat transport character of He II remains considerably different from conventional fluids because this dominant mechanism is internal convection and the loss involves the interaction between the two fluid components.

ACKNOWLEDGEMENTS

Initial construction of the LHFF was supported by the DOE University Research Instrumentation Program under grant DE-FG05-84ER75165. Continuing support for He II experimentation is derived from DOE-High Energy Physics and NASA-Goddard Space Flight Center.

REFERENCES

1)   R. J. Donnelly, presentation this conference.
2)   J. G. Weisend II and S. W. Van Sciver, Adv. Cryo. Engn. <u>33</u>, 507 (1988).
3)   J. G. Weisend II, Y. Huang and S. W. Van Sciver, 7th Intersociety Cryogenic Symposium, Houston, TX, January 22-26, 1989, pp. 141-6.
4)   H. Ito, Trans. ASME, J. Basic Engn. <u>81</u>, 123 (1959).
5)   A. Kashani and S. W. Van Sciver, Adv. Cryo. Engn. <u>31</u>, 489 (1986) and Proc. 11th Intern. Cryo. Engn. Conf., Berlin (1986), pp. 654-8.
6)   J. G. Weisend II, Ph.D. Thesis, University of Wisconsin, Madison (1989) and J. G. Weisend II and S. W. Van Sciver, submitted to 5th AIAA/ASME Thermophysics and Heat Transfer Conference.
7)   L. Dresner, in Cryogenic Properties, Processes and Applications, Vol. 82, AICHE Publication #ISBNO-8169-0303-X (1986), pp. 81-5.
8)   A. Shajii, Y. Huang, M. Daugherty, R. J. Witt and S. W. Van Sciver (accepted for publication in) Adv. Cryo. Engn. <u>35</u> (1990).

A Study of Homogeneous Turbulence in Superfluid Helium

Michael R. Smith and Russell J. Donnelly

Department of Physics, University of Oregon
Eugene, Oregon 97403

Abstract

We generate homogeneous turbulence by means of a moving grid within a stationary channel filled with superfluid helium. The decay in the root mean square vorticity is linked with the quantized line density, which in turn is measured via second sound attenuation. Preliminary results reveal how the time decay of turbulence depends on grid Reynolds number.

1. Introduction and Background

The study of homogeneous isotropic turbulence occupies a unique place in fluid mechanics since it is especially well suited to theoretical investigations, often based upon power spectra and predictions of mean square velocity fluctuations across various scales [1]. Experimentally, such turbulence is generally created behind a grid (at least 65% open) placed within a wind tunnel test section. The turbulence thus generated has a mean velocity, and the flow is examined at various distances behind the grid. In this configuration, it is then straightforward to measure the overall time decay of the mean square velocity fluctuations, $< u^2 >$ at a particular scale. In classical wind tunnel studies [2,3,4] it has been found that $< u^2 > \propto t^{-m}$, where u is a small scale velocity fluctuation and m is a decay exponent dependent upon Reynolds number (here denoted Re_{grid} and based upon mean oncoming velocity and grid mesh length). Experimentally it is found that $m \sim 2.51$ for $1,300 < Re_{grid} < 1,800$ and $m \sim 1.2$ for $17,000 < Re_{grid} < 135,000$. In fully developed turbulence, one can show that [5]

$$< \omega^2 > \sim - \frac{1}{\nu} \frac{\partial < u^2 >}{\partial t} \tag{1}$$

(where $< \omega^2 >$ is the mean square vorticity, a characteristic of the turbulent field, and ν is the kinematic viscosity) so that the mean square vorticity is observed to decay approximately as $t^{-3.51}$ and $t^{-2.2}$ for low and high Reynolds numbers respectively. These decays may be compared with the inviscid [6] and zero Reynolds number [7] solutions in which the mean square vorticity decays as t^{-2} and $t^{-3.5}$ respectively.

In light of current proposals for the use of liquid helium as a test fluid in a modern high Reynolds number turbulence facility (wind tunnel) [8], a logical extension of previous experimental work would be to repeat this work using superfluid helium as the test fluid. This extension offers three advantages over past studies. First, it explores the degree to which superfluid helium mimics classical fluid dynamics at high Reynolds number and small scales. The results may provide clues about phenomena in superfluid helium, such as boundary layer separation and the behavior of large turbulent fields, an understanding of which is critical to the use of superfluid helium as a high Reynolds number test fluid. Second, it is possible to measure the root mean square (rms) vorticity directly in superfluid helium by using second sound absorption. This represents an opportunity to observe first hand something which fluid dynamicists have previously had to infer from velocity measurements. Finally, this experiment may serve as a stepping stone to a better understanding of classical fluid turbulence.

Circulation in superfluid helium is quantized in units of $\kappa = h / m_4$, where h is Plank's constant and m_4 is the mass of the helium atom. Vortices occur in the superfluid component as discrete lines with a core diameter of approximately 2Å . If L is the vortex line length per unit volume, the rms vorticity may be expressed as [9]

$$\omega_{rms} = \frac{1}{2}\kappa L.\qquad(2)$$

Quantized vortex lines attenuate second sound (a temperature wave unique to the superfluid phase) in much the same way as thin wires would attenuate the passage of ordinary (first sound) through air. If a standing wave is set up in a fluid sample between a pair of second sound transducers, the effective vortex line density may be expressed as [9]

$$L = \frac{4\pi\Delta}{\kappa B}\left[\frac{A_0}{A} - 1\right],\qquad(3)$$

where A and A_0 are the received resonant amplitudes with and without vortex lines present, respectively. Δ is the full width at half maximum of the resonant signal, and B is the mutual friction coefficient. Vinen's theory [10] suggests that the free decay of this line density goes as

$$\frac{dL}{dt} = -\chi_2\frac{\hbar}{m_4}L^2,\qquad(4)$$

where χ_2 is a constant of order unity, and $\hbar = h / 2\pi$. Integration yields

$$L = \frac{L_0}{1 + \frac{L_0\chi_2\hbar}{m_4}t},\qquad(5)$$

where L_0 is the intial line density. Note that Equation 3 provides a means of relating the line density (and hence ω_{rms}) to the measured resonant amplitude at a given instant, while Equation 5 provides a possible model for its decay. Theoretically then, one expects this superfluid turbulence to decay as t^{-1} to first order. Milliken, Schwarz, and Smith[11] arrived at a similar decay law by somewhat different physical reasoning. Milliken *et al* generated turbulence between ultrasonic transducers, and their results are in good agreement with the predicted t^{-1} time dependence. However, since there is no clear characteristic length scale or velocity, there does not seem to be any way of associating a Reynolds number with their data for the sake of comparison with classical experiments.

2. Experimental Setup

Our apparatus (see Figure 1) consists of a brass channel whose interior cross section is 1 cm square. It is approximately 60 cm in length, and instrumented along its length with pairs of second sound transducers approximately 5 cm apart. These transducers operate on the same principle as a moving diaphragm type microphone. The vibrating membrane is constructed of nuclepore (filter paper with .1 micron pores) and is transparent to the superfluid component of the liquid helium, thus setting up a fluctuation in the local normal density which propagates across the channel. This propagating variation in the density is formally equivalent to second sound.

The grid was constructed by machining a uniform series of slots, each approximately 1.76mm wide, in a 1.5mm thick square brass wafer, approximately 1cm on a side. The remaining solid portions are approximately 0.6mm wide. We selected the slot configuration so as to minimize the contact atea with the channel around the perimeter of the grid. This in turn minimizes the contribution to the turbulent field due to the shear stress created at the walls. The grid is suspended at the end of a 3/32" rod, and the entire assembly's cross-sectional area is over 65% open . The rod extends outside the bath through a vacuum seal, and is connected to a stepper motor via a cable and drum assembly. The stepper motor allows positional knowlege to well within 0.5mm with our protocol. At 1.65K, this arrangement is capable of achieving grid Reynolds numbers in the range $1.3 \times 10^3 < Re_{grid} < 5.2 \times 10^5$. We thus have a dynamical range extending over nearly three orders of magnitude.

The operating temperature of 1.65K was selected for the local maximum in second sound velocity which occurs there. This local maximum minimizes the effect of temperature fluctuations on our measurements. Temperature is monitored and controlled to within a few μK via one of two 4-wire germanium resistance thermometers in a control loop with a 1000Ω heater. Channel temperature may be monitored using one of two installed 2-wire resistors indicated in the figure. A special film heater is installed in the bottom of the channel for generating counterflow turbulence of the sort studied by Vinen[10], and Swanson and Donnelly[12]. The purpose of this configuration is to compare our results with previously published counterflow data, then extend the data to higher initial line densities and make comparisons with our grid generated turbulence.

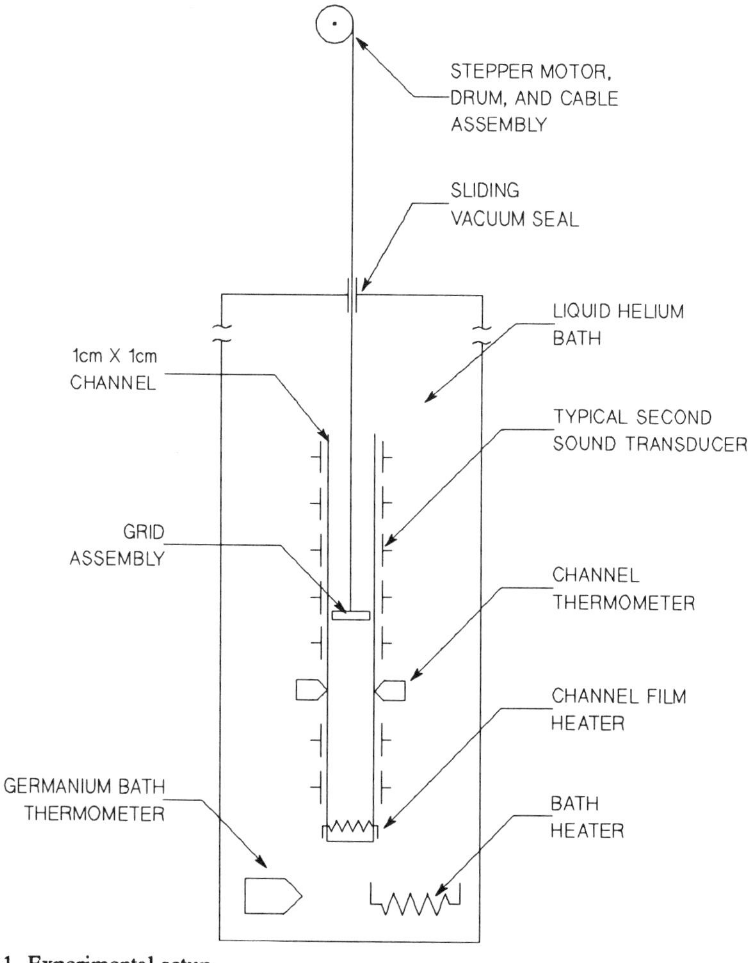

FIG 1. Experimental setup.

3. Preliminary Results and Discussion

In all data sets, $t=0$ is taken to be that point in the raw data (not shown) at which the grid assembly has just cleared the second sound transducer currently in use. This point is located experimentally for each transducer pair by quasi-statically moving the grid up the channel in 0.25mm steps. As long as the grid is below or in front of a particular sensor pair, the rod/grid assembly eclipses the sensor, attenuating second sound transmission across the channel. When the grid is above the sensor pair, second sound is freely transmitted, and attenuation drops to zero. We take the point at which the measured attenuation drops below 10 percent of the maximum eclipsed value to be the instant of grid passage. We find this point to be experimentally repeatable to within 0.25mm when the grid is moved quasi-statically. Stepper motor velocity is somewhat less perfect. We have confirmed that our knowlege of the average velocity is good to within about two percent.

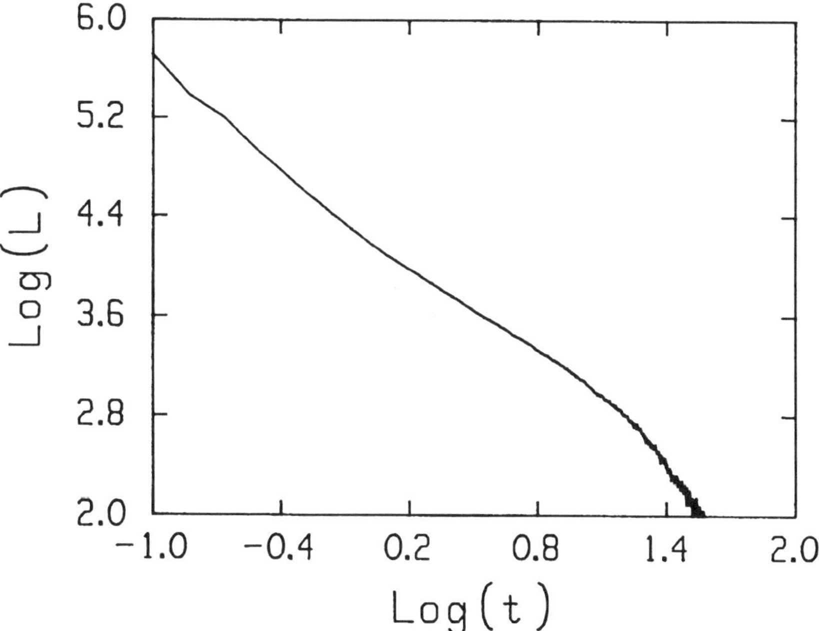

FIG 2. Typical decay profile for line density ($\omega_{rms} \sim L$) associated with low velocity grid generated superfluid turbulence.

A typical decay profile of low velocity grid-generated line density (ω_{rms}) is shown in Figure 2. The goal is to fit a simple decay law of the form $L = t^{-m}$ (recall that $\omega_{rms} \sim L$) to the linear portion on the log-log plot, where m (the decay exponent) is determined from the apparent slope by means of a simple least squares fit. The departures from linearity are not completely understood at this point, but we offer a few plausible explanations. For small times, the deviation from linearity may come from many possible sources. Among these are spatial averaging of the turbulent field (due to the nature of the resonant signal), and departures from strict power law form $L = 1/t^{m}$ (the presence of an additive constant in the denominator along with t^{m} for example). Additionally, the data are convoluted with a response function associated with the resonant second sound wave within the channel. This discrete convolution contributes to rounding

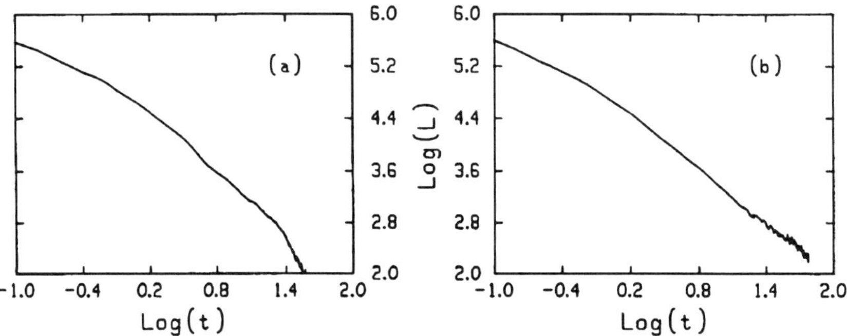

FIG 3. (a): Turbulence decay profile typical of higher grid Reynolds numbers. (b): Clearer decay profile resulting from averaging together data from many runs at the same Reynolds number.

at small time, and also tends to make the slope slightly more pronounced. Though computer simulations indicate that this effect small, it will be helpful to employ discrete de-convolution techniques on future data. Discrete time convolution also seems to introduce small fluctuations in the signal whose effect may become quite pronounced as the signal approaches its final zero turbulence value. Finally, many decay profiles show a marked transition from initial power law dependence as the line density drops below approximately 10^{3}. This transition, characterized by a more pronounced slope is not clearly understood at this time, but may have its analogy in classical fluids, where similar behavior is attributed to the possibility that the turbulent field has left the so-called self-preserving state.

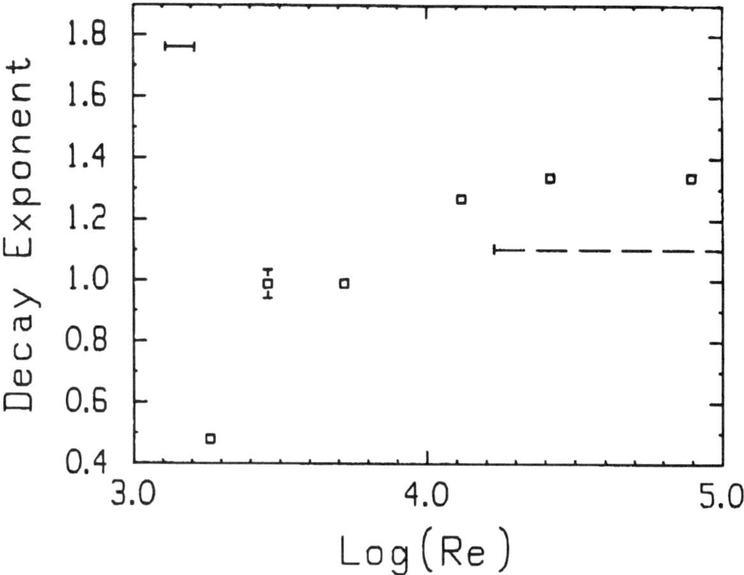

FIG 4. Decay exponent m vs generating Reynolds number, where a simple decay law of the form t^{-m} has been fitted to the data. Horizontal dashed lines indicate classical results[3,4].

We sought to minimize the importance of small fluctuations (due to noise in the signal, possible non-uniformities in the tangle of vortex lines, or time convolution effects), by averaging an ensemble of many runs for a given grid velocity on a point by point basis (all points are equally spaced in time). Figures 3a and 3b allow comparison of data before and after such averaging. Runs were averaged in this manner over a wide range of Reynolds numbers, and slopes (decay exponents) were derived from the averaged profiles by the least squares fit method mentioned above, keeping in mind that each point in an averaged profile now represents an average over the ensemble at that instant in time, with its own standard deviation. The results are shown in Figure 4, where error bars are statistically derived. One striking feature of this plot is the region of fairly low Reynolds numbers $(2,800 < Re_{grid} < 7,900)$ in which the decay exponent is close to 1, and thus is in good agreement with the inviscid solution ($< \omega^2 > \propto t^{-2}$ implies $\omega_{rms} \propto L \propto t^{-1}$). There is also agreement in some sense with the results of Milliken et al[11]. Though there is no clear way of determining any equivalent grid Reynolds number for their work (to aid in comparing with our results), the maximum initial line densities which they generated are consistent with our extrapolated initial line densities over the range in which the decay exponent was close to 1.1.

If superfluid helium behaved as a purely classical fluid, one might expect (based on wind tunnel experiments [3,4]) the line density to decay as $L \propto t^{-1.76}$ for $1,300 < Re_{grid} < 1,800$ and $L \propto t^{-1.1}$ for $17,000 < Re_{grid} < 135,000$. These data are plotted as horizontal dashed lines in figure 4. Observe that experimentally, classical homogeneous turbulence approaches the inviscid solution at higher Reynolds numbers, while our results seem to exhibit this behavior in the low-to-mid Reynolds number range. This agrees with our understanding of superfluid helium dynamics, where we expect viscosity effects to be negligible at low velocities, and to play an increasing role as the velocity approaches the speed of second sound. In a classical fluid however, viscosity is not velocity dependent, so that terms in the Navier-Stokes equations which are proportional to viscosity (linear in velocity) are permitted to play a less important role (in comparison with velocity squared terms, for example) at higher velocities. Thus many classical flows approach the inviscid solution at higher Reynolds numbers. Classical and superfluid experimental results seem to be in only general agreement over the approximate range $7,900 \leq Re_{grid} \leq 78,000$.

The cause of the discontinuity at very low Reynolds numbers is not completely understood at this point. If the two fluids are locked and flowing together over much of the curve, there must be some critical Reynolds number below which they unlock (though they would of course remain coupled through the interaction of the vortex lines). Beyond this, very little can be said.

What is not clear from figure 4 is the nature of the transition from one value of m to another as the Reynolds number is varied. In looking at the set of averaged decay profiles (where the standard deviations of each point in the linear regions are remarkably small), one notes that the value of m doesn't seem to vary smoothly in the traditional sense from one Reynolds number to the next. Instead, there may be two or more linear regions. One of these is often clearly dominant, providing the better fit, while another smaller linear region may exibit a slope more characteristic of a neighboring Reynolds number. As the Reynolds number is varied, the data suggests that one such region may be growing, as others shrink.

4. Summary and Look Ahead

In summary, the rms vorticity in superfluid helium appears to decay as an inviscid fluid at low Reynolds number, while the decay exponent m in the decay power law $L \propto t^{-m}$ exhibits some Reynolds number dependence at higher values of Re_{grid}. Furthermore, m does not appear to vary smoothly with Reynolds number in the traditional sense. The agreement between these results and those derived from classical wind tunnel experiments is somewhat less than perfect.

There are several refinements currently being implemented. First, it should be possible to deconvolve the data from the response time of the system. This will improve our knowledge of grid location and thus reduce possible errors in absolute determination of the value of m, as well as its functional dependence on Re_{grid}. Second, we are currently modifying the grid to further reduce the anomolous contribution of shear stress originating at the walls to the turbulent field. Looking ahead, it will be interesting to create counterflow turbulence by generating bursts of heat with the special film heater at the bottom of the channel. The decay of this counterflow turbulence will then be compared with that of grid generated turbulence, hopefully providing new and useful insights into the dynamics underlying macroscopic superfluid turbulence.

Finally, the most exciting prospect for the future seems to lay in installing a superfluid Fountain Effect Pump (FEP) at one end of the channel, in effect creating a pseudo-classical tunnel capable of continuous operation at the same Reynolds numbers discussed above. Such an endeavor would not only serve as a further point of comparison with classical wind-tunnel studies of grid turbulence, but would enable us to take time series data, and even study wakes behind common objects such cylinders (whose large scale wake structures are well documented in classical wind-tunnels). Such studies would not only contribute uniquely to our knowlege of liquid helium phenomenology, but may actually enhance our knowlege of classical vortex fields in general.

This research was sponsored by the Defense Advanced Research Projects Agency through a grant from the United States Office of Naval Research, N00014-89-J-1274. Facilities and equipment were also provided by the National Science Foundation low-temperature physics program.

5. References

[1] see for example Batchelor,G.K.,1953, The Theory of Homogeneous Turbulence. Cambridge University Press, or Proudman,I., and Reid,W.H.,1954, *On the decay of a normally distributed and homogeneous turbulent velocity field*. Philos. Trans. R. Soc. London Ser. A **247**,1643-89.

[2] Gence,J.N.,1983, *Homogeneous turbulence*. Ann. Rev. Fluid Mech. **15**,201-22.

[3] Comte-Bellot,G., and Corrsin,S.,1966, *The use of a contraction to improve the isotropy of grid generated turbulence*. J. Fluid Mech. **25**,657-82.

[4] Bennet,J.C., and Corrsin,S.,1978, *Small Reynolds number nearly isotropic turbulence in a straight duct and a contraction*. Phys. Fluids **21**,2129-40.

[5] see for example sections 16 and 34 of Landau,L.D., and Lifshitz,E.M., 1987, Fluid Mechanics, second edition, Pergamon Press.

[6] Proudman,L., and Reid,W.,1954, *On the decay of a normally distributed and homogeneous turbulent velocity field*. Philos. Trans. R. Soc. London Ser. A **247**,163-89.

[7] von Karman,T., and Howarth,L.,1938, *On the Statistical Theory of Isotropic Turbulence*. Proc. R. Soc. London, Ser. A **164**,192-215.

[8] Donnelly,Russell, Smith,Michael, and Swanson,Charles, 1990, *Superfluid Wind Tunnels*, Physics World **3**,39-41.

[9] Swanson,Charles E.,1985, A study of vortex dynamics in counterflowing helium II, Ph.D. Thesis,University of Oregon.

[10] Vinen,W.F.,1957, *Mutual friction in a heat current in liquid helium II. III. Theory of mutual friction*. Proc. R. Soc. London, Ser. A **242**,493-515.

[11] Milliken,F.P.,Schwarz,K.W., and Smith,C.W.,1982, *Free decay of superfluid turbulence*. Phys. Rev. Lett. **48**,1204-7.

[12] Donnelly,Russell J., and Swanson,Charles E.,1986, *Quantum turbulence*. J. Fluid Mech. **173**,387-429.

Thermal Convection in Liquid Helium

Joseph J. Niemela and Russell J. Donnelly
Department of Physics, University of Oregon, Eugene, Oregon 97403

Abstract

Above about 2.2 K liquid helium is assumed to behave as an ordinary incompressible Newtonian fluid, free from the macroscopic quantum effects possessed by its lower temperature counterpart. Recent thermal convection experiments support this assumption.

Helium is remarkable in the sense that it remains liquid under its own vapor pressure to the absolute zero of temperature. Below about 2.2 kelvin the liquid phase (helium II) can be described by a two-fluid model which supposes two interpenetrating fluids - an inviscid "superfluid" component and a "normal" component. Above 2.2 kelvin, on the other hand, it is assumed that the fluid (helium I) differs from ordinary Newtonian fluids only in temperature, and that its behavior should therefore be correctly described by the Navier-Stokes equations for a single fluid. One suitable test of this assumption is to compare the observed onset and nature of hydrodynamic instabilities in helium I with detailed predictions based on the stability equations derived for ordinary incompressible fluids.

A particularly simple and experimentally accessible hydrodynamic system is Rayleigh-Benard (RB) convection, in which a thin horizontal layer of fluid is uniformly heated from below. For vertical temperature differences below some critical value, which depends on geometrical factors and the properties of the particular fluid, the fluid remains motionless, while above it the fluid undergoes a continuous transition to convective flow. We will consider a special case in which the heating is periodically varied in time, a subject of current interest both theoretically and experimentally, and for which a number of novel states of the system can be investigated via tuning of the appropriate external parameters.

In order to have a more universal measure of the state of the RB system, the vertical temperature difference is usually made nondimensional in terms of the Rayleigh number $Ra = g\alpha\Delta T d^3/\nu\kappa$, where g is the gravitational acceleration, α, ν, and κ are respectively the isobaric thermal expansion coefficient, kinematic viscosity, and thermal diffusivity of the fluid, and ΔT is the vertical temperature difference across the fluid layer of height d. In practice, the finite geometry of the experimental cell, as well as imperfect thermal boundary conditions, have some influence on the correspondence between a particular value of the Rayleigh number and the state of stress of the system, so that a more accurate representation of the latter is via a *reduced* Rayleigh number $\epsilon = Ra/Ra_c - 1$, where Ra_c is the critical value of the Rayleigh number determined at the onset of convection. In addition, particular fluids may be characterized by the ratio of kinematic viscosity to thermal diffusivity, or Prandtl number $\sigma = \nu/\kappa$. Helium I has a Prandtl number varying with temperature from slightly less than $1/2$ to about $3/4$, where by comparison, σ is about $2/3$ for ideal gases, 6 for water, and about 200 for silicone oils.

The use of helium I as a medium for studying convective flows has been recently reviewed by Behringer[1]. There are a number of technological advantages to using helium I as a working fluid, such as the existence of a well-developed cryogenic technology for precise temperature measurement and control. Another advantage is the large ratio of thermal diffusivities of the solid (copper) horizontal boundary plates to the enclosed fluid at very low absolute temperatures. This enables fast and virtually unattenuated reception of high-frequency temperature fluctuations, an extremely useful attribute, for instance, in the study of time-periodic or turbulent convective flow. Conversely, one can also easily *apply* a time-dependent fluctuation of the temperature to the fluid. This suggests a class of experiments which was mentioned above; namely, the temporal modulation of the bouyancy force which drives the convective flow. This area of investigation was originally motivated, at least in part, by similarities to the well-known problem of the inverted physical pendulum[2] which can be made to execute stable oscillations in an *upright* position via vertical modulation of its point of support. In the hydrodynamic analogue, enhanced stability is achieved by allowing the temperature difference across a layer of fluid to possess both a steady mean value and an oscillating component such that $\Delta T(t) = \Delta T\{1 + \Delta\cos(\omega t)\}$. The reduced Rayleigh number for this problem then is defined as $\epsilon = Ra / Ra_c^{STAT} - 1$, where Ra_c^{STAT} is the critical value of the Rayleigh number in the absence of modulation, and Ra is here taken to be averaged over the modulation cycle. In this system the conducting state of the fluid for a range of mean Rayleigh numbers *greater* than Ra_c^{STAT} is made stable by this periodic modulation of the instantaneous temperature difference. That is, while standard RB convection is initiated, by definition, at $\epsilon_c = 0$, the convective threshold with modulation is shifted to higher values of the control parameter, $\epsilon_c > 0$.

In our experiments the helium was confined in a cylindrical container consisting of copper top and bottom plates spaced a distance d = 0.0917 cm apart. The sidewalls of the cell were made of thin (0.015-cm thickness) stainless steel and the cell had a radius-to-height ratio of about Γ = 8.82. Temperatures were measured with germanium resistance thermometers embedded in the top and bottom plates. For standard convection, the top plate temperature was held fixed by servo-ing the electrical current to an attached resistance heater, while a superfluid helium bath served as a heat sink, connected to the top plate via a copper braid. A constant heat current was applied to the bottom plate to induce a temperature difference across the helium layer, with a relatively small portion of the heat being conducted in parallel through the thin stainless steel sidewalls. The entire cell assembly was located in an enclosed volume which was evacuated to prevent convective heat transfer between the plates outside the sidewalls. The mean temperature of the top plate was primarily held at 2.63 K, corresponding to a Prandtl number σ = 0.49.

To modulate the temperature difference across the helium I layer required periodically heating and cooling the top plate of the cell, while simultaneously applying a steady heat current to the bottom plate. Evidence of the convective transistion was obtained by calculating the Nusselt number Nu from heat flux measurements, where Nu is defined as the ratio of the effective thermal conductivity of the fluid to that due to conductive heat transfer alone. In figure 1(a) we show Nusselt numbers as a function of ϵ for various amplitudes Δ of the modulation, and at a fixed value of the modulation frequency $\tilde{\omega} = \omega \tau_v = 10.7$, where $\tau_v = d^2 / v$ represents a characteristic vertical thermal diffusion time for the fluid layer. An increase in heat transfer from the conduction value Nu = 1 indicates the onset of convective motions and it is clear from figure 1(a) that the modulation does, indeed, have a stabilizing effect on the conducting state. Ahlers, Hohenberg, and Lucke[3] have made quantitative calculations of the threshold shifts as a function of σ, Δ, and ω. In figure 1(b) we show their theoretical results for $\tilde{\omega} = 10.7$ and $\sigma = 0.49$ (solid curve), together with the experimentally determined shifts from the data in figure 1(a). The abscissa is marked in units of a complex, frequency dependent amplitude $\Delta'(\omega)$ which takes into account the fact that the temperature profile in the conduction state becomes increasingly nonlinear at high frequencies of modulation (see ref[3]). It is clear that theory (derived for classical fluids) and experiment are here in excellent agreement. A closely related system which exhibits qualitatively very different behavior involves high frequency, large amplitude temperature modulation about *zero* mean. By this we mean that the temperature difference between the plates has the form $\Delta T = T_o \cos(\omega t)$, where the period of modulation is small compared to some appropriate physical time scale. This rapid modulation forms a thermal boundary layer, or "Stokes" layer, near the plate of thickness $\delta = (2\kappa / \omega)^{1/2}$. Gershuni and Zhukhovitskii[4] (GZ) showed that this Stokes layer would lose stability to convection as a critical modulation *amplitude* was exceeded, despite the absence of any mean temperature gradient across the fluid. They defined a modified Rayleigh number, with δ replacing the fixed cell height d and T_o replacing ΔT: $Ra = g\alpha T_o \delta^3 / v\kappa$. In analogy to standard RB convection this

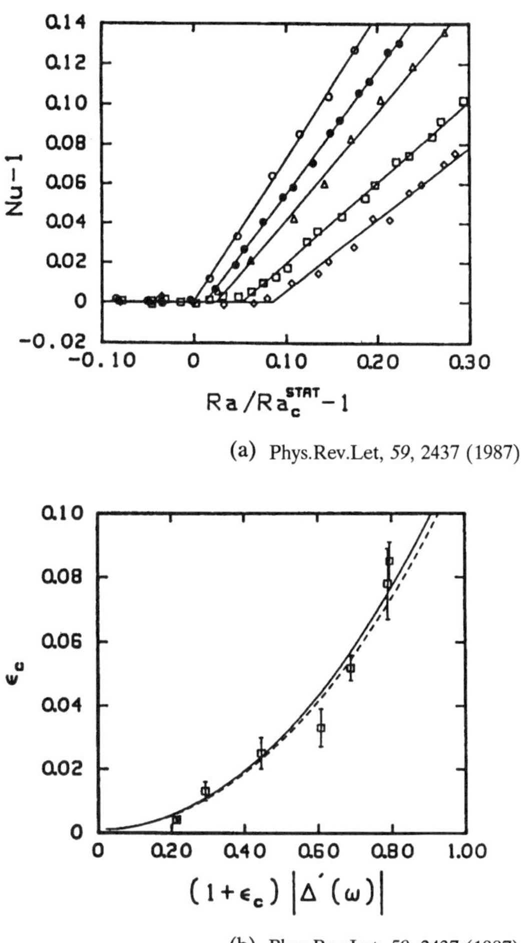

(a) Phys.Rev.Let, *59*, 2437 (1987)

(b) Phys.Rev.Let, *59*, 2437 (1987)

Figure 1. (a) The effect of changing amplitude at a fixed frequency: $\omega = 10.7$. Open circles, $\Delta=0$; solid circles, $\Delta=0.30$; triangles, $\Delta=0.45$; squares, $\Delta=0.68$; diamonds, $\Delta=0.76$. The sloping solid lines are least-squares fits to the data. (b) Convective threshold as a function of amplitude for data of figure 1(a). The solid and dashed lines represents stability results obtained from ref [3]. The solid line is for the ideal system, while the dashed line represents the effects of convective forcing in the real system due to imperfections.

Rayleigh number has a unique critical value (for a given Prandtl number) so that the actual critical modulation amplitudes scale with the boundary layer thickness in the same manner as critical temperature differences scale with the fixed cell height under steady heating conditions. Swift and Hohenberg[5] (SH) have extended the linear theory of GZ to the nonlinear regime and predict both qualitative and quantitative differences in the convective transition; for instance, the normally forward bifurcation is predicted to become inverted, or hysteretic.

Experimentally, this high-frequency modulation was carried out in much the same manner as before, with the top plate of the cell periodically heated and cooled. However, in this case there was no steady heat current applied to the bottom plate, and in addition the cell height was considerably larger (d = 0.4115 cm) so that the condition $\delta << d$ was easily satisfied, which ensured that the bottom plate did not interfere with the formation of the boundary layer. This is the sense in which the modulation is termed "rapid" above. Detection of the convective flow was as follows. In the absence of convection, periodic heating and cooling of the fluid is completely reversible, i.e., there cannot be any net vertical heat flux over a cycle of the modulation. Therefore, the bottom plate will possess precisely the same *mean* temperature as the top plate. With convection, on the other hand, the vertical heat flux becomes a superposition of the conductive heat transfer (which is reversible) and that due to the convective motions. These motions are maintained by a *net* transfer of heat, which cools the bottom plate and gives rise to a time-averaged temperature difference $\Delta T = T_b - T_t < 0$, where T_b and T_t are the mean temperatures of the bottom and top plates respectively.

Figure 2(a) shows the measured temperature difference across the cell, normalized by the frequency-dependent critical modulation amplitude T_{oc}, as a function of Ra for $\sigma = 0.49$ and three different modulation frequencies, $\omega / 2\pi = 0.032$, 0.048, and 0.080 Hz. It is clear that convective instability occurs at a common Rayleigh number for each run, and

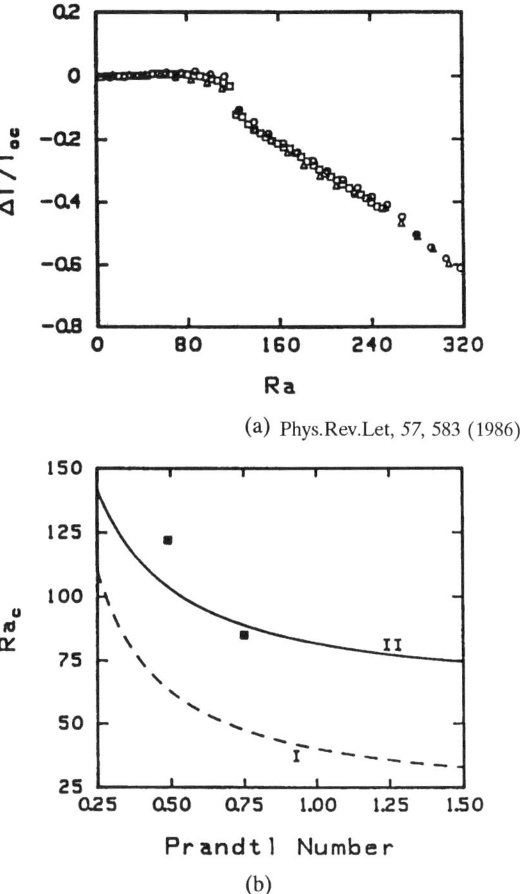

(a) Phys.Rev.Let, *57*, 583 (1986).

(b)

Figure 2. (a) Dimensionless temperature difference vs. Stokes layer Rayleigh number for $\sigma = 0.49$ and various modulation frequencies: circles, 0.032 Hz; squares, 0.048 Hz; triangles, 0.08 Hz. (b) Critical Rayleigh number as a function of Prandtl number σ. The lines are predicted values from ref [4], where the solid line (II) corresponds to rigid boundaries, as in the experiment, while the dashed line is for stress-free boundaries. Solid squares are experimental results for $\sigma = 0.49$ and 0.75.

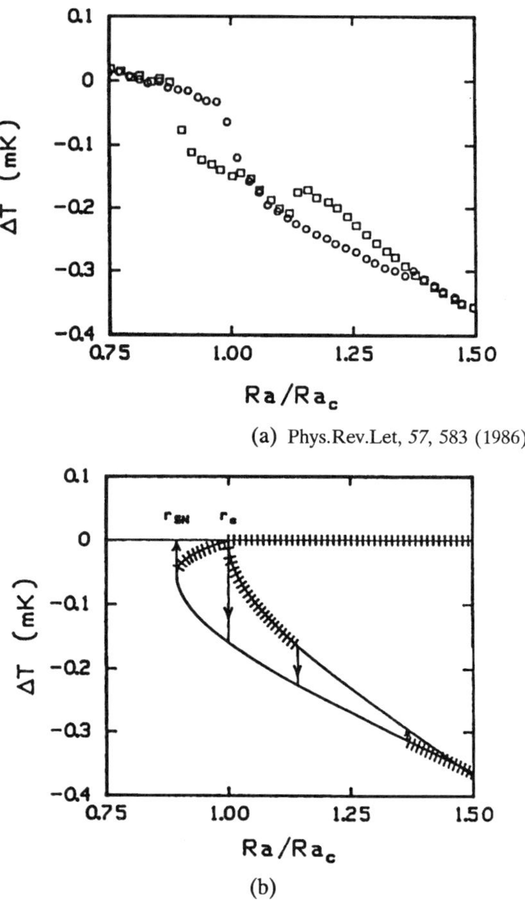

(a) Phys.Rev.Let, *57*, 583 (1986).

(b)

Figure 3. (a) Hysteresis data for $\sigma = 0.49$ and a modulation frequency of 0.048 Hz. Circles are for increasing Rayleigh numbers, while squares represent decreasing Rayleigh numbers. (b) Schematic bifurcation of the hysteresis. The unhatched portions of the curves below $\Delta T = 0$ are plotted to overlay the data of figure 3(a), while the unhatched portions are extrapolations suggested by theoretical plots of a similar nature. The arrows show the direction of the transitions observed. The critical Rayleigh number is denoted by r_c while the saddle node bifurcation is labeled as r_{SN}.

we show in figure 2(b) a comparison of this *absolute* number, as well as the critical Rayleigh number for $\sigma = 0.75$, with the calculations of GZ. Note that the transition is clearly *not* continuous, as it was for lower frequency modulation, in agreement with the predictions of SH. To obtain information about the hysteretic nature of the bifurcation, we ramped the Rayleigh number in both directions through the transition as shown in figure 3(a). A schematic view of this data is illustrated in figure 3(b), which shows clearly both the temperature drop at onset, and the size of the

subcritical region ($Ra < Ra_c$). The saddle node, where the system returns to the conducting state upon decreasing Ra, occurs in the experiment for $\epsilon = -0.14 \pm 0.05$, which agrees extremely well with the theoretical value $\epsilon = -0.15$ predicted by SH. They also predict a value of the normalized temperature drop at onset of $\Delta T / T_{oc} = 0.05$. It is difficult to measure this quantity accurately in the experiment due to the presence of significant convective forcing close to the onset (as evidenced by the considerable rounding near $\epsilon = 0$), but from figure 2(a) we estimate $\Delta T / T_{oc} \approx 0.07$.

In conclusion, we find good agreement, both qualitatively and quantitatively, between "classical" theory and experiment for modulated driving of convective flow in helium I. This provides further evidence that above 2.2 kelvin liquid helium can be considered a Newtonian fluid whose behavior is correctly described by the single-fluid Navier-Stokes equations.

References

1. Behringer, R.P., 1985 "Rayleigh-Benard convection and turbulence in liquid helium." Rev. of Mod. Phys. **57**.

2. Landau, L.D., & Lifshitz, E.M., 1960 *Course of Theoretical Physics. Vol. 1. Mechanics.* (Pergamon Press).

3. Ahlers, G., Hohenberg, P.C., & Lucke, M., 1985 "Thermal convection under external modulation of the driving force. Part 1. The Lorenz model." Phys. Rev. **32A**, 3493-3518.

4. Gershuni, G.Z., & Zhukhovitskii, E.M., 1976 *Convective instability of incompressible fluids.* Keter, Jerusalem, 218-222.

5. Swift, J.B., & Hohenberg, P.C., 1987 "Modulated convection at high frequencies and large modulation amplitudes." Phys. Rev. **36A**, 4870.

Helium Fluid Flow Facility Cryogenic System

G. E. McIntosh and K. R. Leonard

Cryogenic Technical Services, Inc., Boulder, Colorado

1. Introduction

Either supercritical normal or pressurized superfluid helium flow tunnels are demanding exercises in cryogenic technology. Design requirements include low cryostat heat leaks similar to low loss storage dewars, heat exchangers with low pressure drop and temperature difference, take-apart contruction which is leak-free and convenient, magnetic suspension and balance systems (MSBS) for model positioning and force data collection, model viewing through multiple shells, interchange of models without system warm-up, and operation that is safe and relatively easy to control. Added to these criteria are fundamental choices of He I or He II as the working fluid and design of a "wet" tunnel surrounded by a helium bath or a "dry" tunnel with an internal heat exchanger.

Our purpose here is to define a typical cryogenic fluid flow facility and examine the major design elements. This description is intended to establish the basis for future design work and the outline for a system proposal. Because superfluid helium provides the most dramatic fluid flow characteristics, the discussion is keyed to a He II system operating at 1.8 K and approximately atmospheric pressure. Principal elements described in following sections include the system schematic, He II cryostat, refrigeration requirements, MSBS installation, and a model exchange mechanism.

2. He II Cryogenic System

Details of a He II cryogenic flow system are shown in Figure 1. This is a "wet" system with the flow loop entirely surrounded by one atmosphere helium at 1.8 K. The advantage of this arrangement is that the entire tunnel surface serves as a heat exchanger and there are no internal flow restrictions related to thermal requirements. Although more helium is required, the tunnel fabrication is simplified in a wet configuration because only nominal leak-tightness is required.

Elements of this facility include a normal helium supply dewar, the cryostat, internal cooling system, vacuum pumps and helium recovery gas bags, controls and safety devices, and the MSBS which is not shown. The cryostat consists of a vacuum jacket, an insulated thermal shield cooled by vent vapor, and the helium containment vessel. Internal cooling is provided by a subcooler, a J-T throttle valve, a low pressure (1500 Pa) liquid reservoir, and a superfluid heat exchanger. Principal controls and safety devices for the cryostat are shown on the schematic. It is likely that more will be added as the design is refined.

The arrangement shown assumes an operating heat load which is greater than the capacity of the refrigerator/liquefier. In this case, the liquefier is operated until a quantity of helium is accumulated in the supply dewar before the start of flow test. During the test, refrigeration is produced by flashing subcooled normal helium to approximately 1500 Pa by means of external vacuum pumps. Additional transient refrigeration can be made available by utilizing enthalpy of the superfluid bath and flow loop. Enthalpy change from 1.6 to 1.9 K at one atmosphere is 844.5 j/kg_l (124.3 j/litre). Thus, a 1,000 litre volume can support a 1 kW heat load for some two minutes while warming from 1.6 to 1.9 K without any other refrigeration. Finally, sizing the refrigerator to match the flow test load eliminates the supply dewar and allows cold vapor to return directly to the cold box from the subcooler.

3. He II Cryostat

A simplified cryostat concept is shown in Figure 2. It consists of the inner liquid helium containment vessel, thermal radiation shield, the vacuum enclosure, and a model access stack. The refrigeration subcooler is shown housed in the stack assembly to preserve close shell clearances and thereby minimize overall size of the apparatus. Aside from the aluminum shield, shell selection is a matter of cost and operating convenience. A 8083 aluminum inner vessel and indium-sealed closure flange is practical if the number of penetrations and windows is limited. Stainless steel may be more reliable if there are a large number of complex penetrations. Aluminum is probably the economic choice for the vacuum jacket.

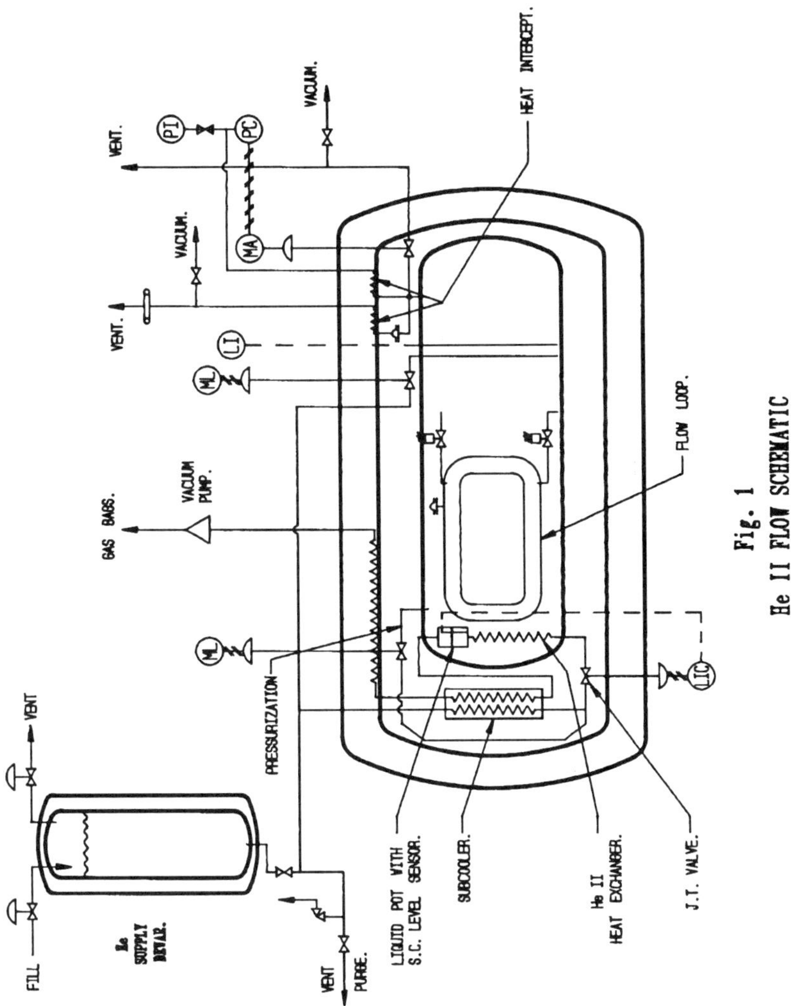

Fig. 1
He II FLOW SCHEMATIC

Thermal performance of the cryostat is based on helium dewar technology. Low heat transfer between the shield and inner vessel is achieved by using low emissivity, less than 3 %, surfaces without multilayer (MLI) insulation. In addition to lower heat leak than an effective thickness of MLI,the combination of high vacuum plus low emissivity surfaces is quicker to cool down and much easier to re-evacuate after the cryostat is opened. The cryostat departs from dewar practice in having only one vaporcooled shield with a 3 to 4 cm blanket of MLI between it and the vacuum jacket. Heat leak caused by the stack, instrumentation, and various electrical leads produces sufficient vapor to cool the thermal shield below 90 K under static conditions. This obviates the need for additional shields with their attendant complexity and space requirement. Much higher vapor generation during fluid flow testing will drop the shield temperature substantially with a significant reduction in heat leak.Because the cryostat is for static use and requires only modest seismic capability, the support system can be designed for a low (2 or 3) multiple of normal forces. Use of long, heat stationed fiberglass/epoxy straps at each end of the cryostat will contribute 5 to 10 % to the total heat leak. Overall, the projected heat leak of the basic cryostat is around 2 watts with an undetermined additional static load of several watts for the flow loop
components.

4. Model Exchange and Withdrawal

Utilization of a cryogenic fluid flow facility is greatly enhanced by the ability to exchange models without having to warm up and open the cryostat. The mechanics of a model exchange system are particularly onerous in a pressurized He II tunnel which has no free liquid surface. This is a challenging design problem for which a workable solution has not been obtained. Thus, Figures 3 and 4 are presented only as conceptual steps in defining the problem.

Figure 3 is an expanded view of the stack shown in Figure 2. The concept illustrated is a three-chamber conditioning system with fluid and vacuum-tight closures between volumes. Chamber operating temperatures are ambient to 90 K in the top, 90 to 4.2 K in the middle, and 4.2 to 1.8 K in the lower chamber. The conceptual operating sequence is to lower the model from one chamber to another by three pairs of positioning hooks with thermal conditioning at each stage. In the final step, an electromechanical door in the top of the tunnel is opened and the model is placed on retractable rests in the test section. Actual positioning of the model in the test section is accomplished by the MSBS.

Figure 4 shows the concept design of a rectangular gate valve which might be used to separate the model loading chambers. This valve combines the wedging linkage typical of large vacuum gate valves and a lip seal design which was successfully used on a superfluid helium bayonet project[2]. Thus, it appears that there are feasible mechanical solutions to the model exchange problem. The more difficult part is to combine workable mechanics with a low heat leak design. Much work is required in this area in order to develop an overall satisfactory model exchange system.

5. Fluid Flow Tunnel Refrigeration

Refrigeration requirements for the cryogenic fluid flow tunnel fall into three categories: static heat leak of the cryostat, standby heat load with the MSBS magnet leads connected, and the operating load of the facility. The static heat leak of the cryostat is equivalent to about twice that of a well designed helium dewar of similar size. Standby heat leak will be higher than the static load because the zero current loss due to vapor-cooled leads is 60 % of full load. The zero current heat load (to normal helium) is about 1 watt for a minimum of eight MSBS magnets designed for 100 amp each. This increases to 1.6 W at full current. Other electrical and instrumentation leads connected during standby will also contribute to the increased thermal load. Finally, the tunnel will dissipate from 100 to 300 W during flow tests depending on the Reynolds number and, possibly, much higher rates for short periods. These loads are summarized as less than 5 W static, 5 to 10 W on standby, and 100 to as high as 1,000 W for flow testing.

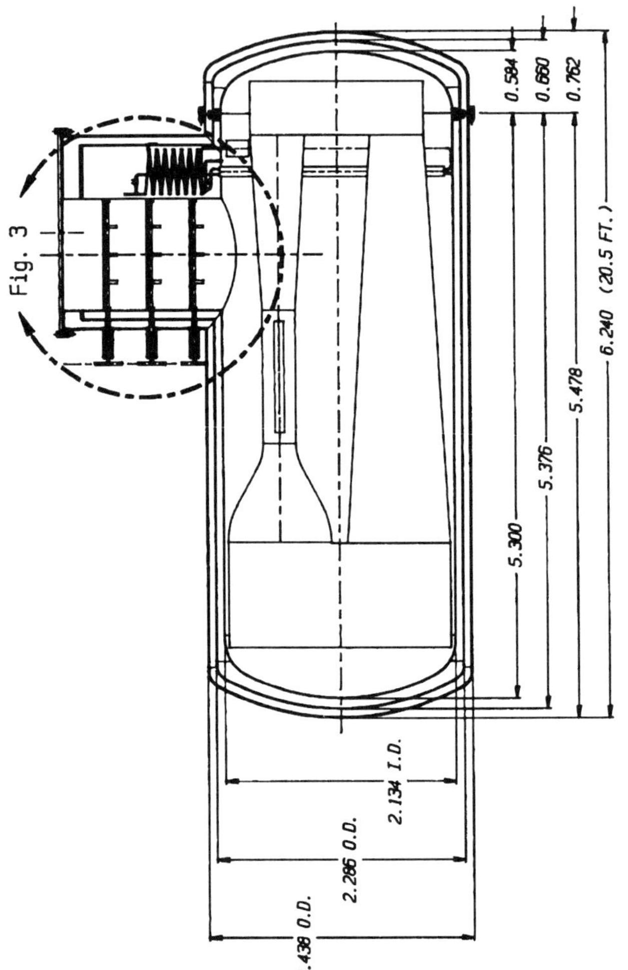

Fig. 2
FLUID FLOW SYSTEM
TYPICAL CRYOSTAT SHELL ARRANGEMENT

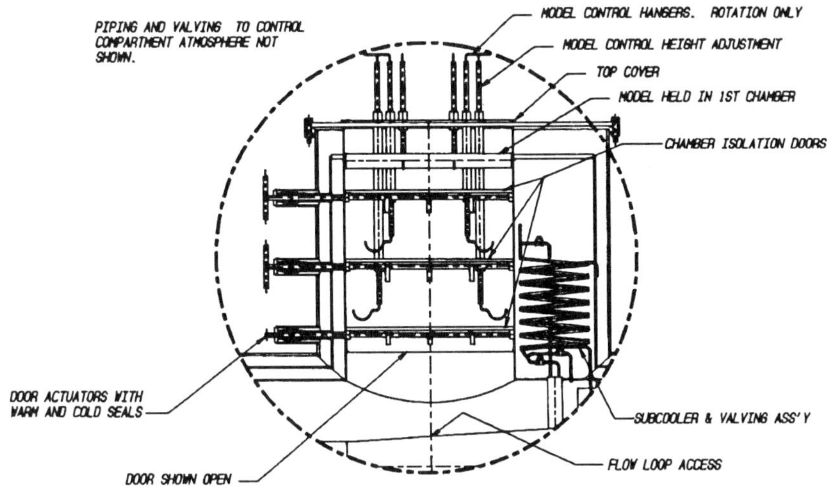

Fig. 3

FLUID FLOW SYSTEM

MODEL EXCHANGE STACK CONCEPT

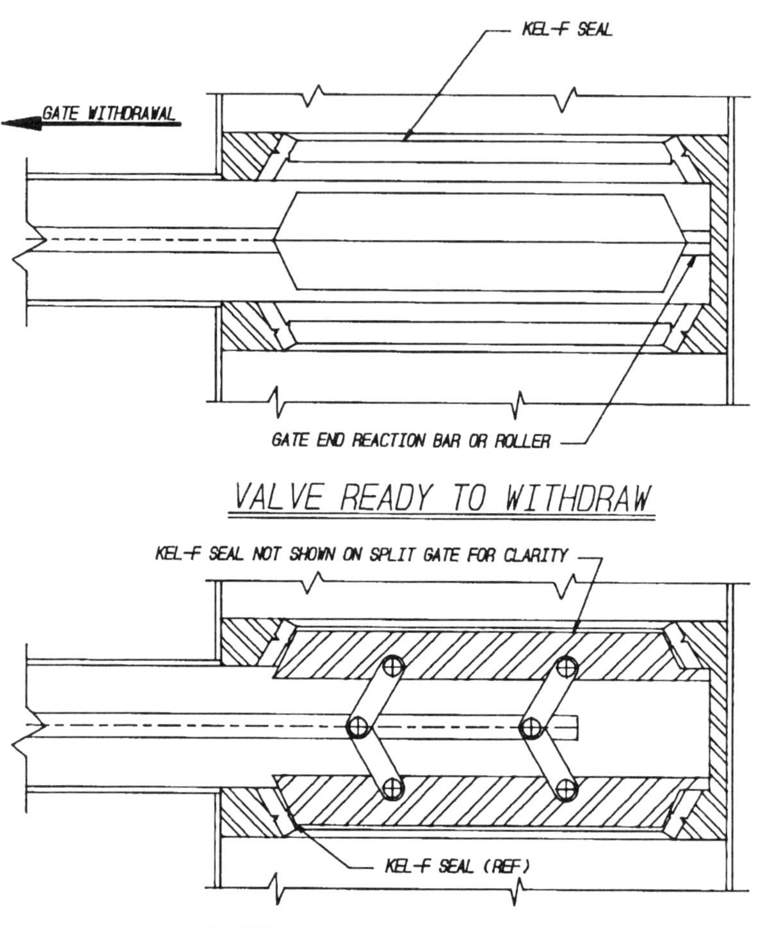

Fig. 4
MODEL WITHDRAWAL GATE VALVE

The disparity in refrigeration requirements from static to flow testing immediately suggests some kind of thermal flywheel arrangement. This is best accomplished with a liquefier, storage dewar, vacuum pump, and gas recovery bag combination. The liquefier is sized to reliquefy the static loss of the system and accumulate liquid in the dewar at a rate sufficient to support the test program. The storage dewar must be large enough to store all the liquid needed for a periodic test plus a 20 to 25 % margin. The storage quantity may include the amount of helium needed to cool the cryostat from 4.2 to 1.8 K if testing is infrequent. The vacuum pump or pumps must have sufficient capacity, about 0.05 g/s per watt at 1.8 K, to hold the tunnel at constant temperature during flow tests at design Reynolds number. (Transient high Reynolds number tests should utilize stored enthalpy of fluid in the tunnel and reservoir.) Finally, the gas recovery bags are sized to hold the excess helium gas produced in a test cycle.

6. Magnetic Suspension and Balance System

The cryogenic fluid flow tunnel is an ideal application for a MSBS utilizing superconducting magnets. There is a ready-made helium reservoir for the magnets which eliminates the need for a separate cryostat system and the forces are relatively low so that rather small magnets are sufficient. MSBS technology is becoming mature[3,4] so that design of the magnets and control system is straight- forward. The position sensing system may be the most difficult part of the MSBS. Choices include operating the light sources and receivers in liquid helium, use of fiber optics arrays, or building the tunnel and cryostat with four sizeable windows for external viewing and control. A fiber optic system currently appears most attractive.

7. Conclusions

Technology exists to design and fabricate a feasible cryogenic fluid flow tunnel utilizing either normal helium or pressurized He II. Conventional techniques are adequate for the helium storage dewar, refrigerator/liquefier, cryostat, and the He II cooling system. Special design attention is required for the MSBS, cold model exchange mechanism, and the MSBS position sensors.

8. References

1. V. D. Arp and R. D. McCarty, "Thermophysical Properties of Helium-4 from 0.8 to 1500 K with Pressures to 2000 MPa," <u>NIST Technical Note 1334</u>, U. S. Dept. of Commerce (1989).

2. G. E. McIntosh et al, Bayonet for Superfluid Helium Transfer in Space, in: "Advances in Cryogenic Engineering," Vol. 33, Plenum Press, New York (1988), p. 885.

3. R. W. Boom et al, Magnetic Suspension and Balance System Advanced Study--Phase II, <u>NASA Contractor Report 4327,</u> NASA (1990).

4. C. Britcher et al, A Flying Superconducting Magnet and Cryostat   for Magnetic Suspension of Wind-Tunnel Models, in: <u>Cryogenics,</u> Vol. 23, No. 10, October 1983.

Appendix

Power economy in high-speed wind tunnels

by choice of working fluid and temperature

R. Smelt, M.A.

SUMMARY

The power required to operate a high-speed wind tunnel at fixed Mach number, Reynolds number, and pressure can be greatly reduced if instead of air at normal temperatures, other fluids or low temperatures are employed.

If operation at normal temperatures is desired, best power economy is obtained by using certain fluorine compounds of high molecular weight. Of these, certain hexa-fluorides (S F_6, SeF_6, Te F_6) take first place; the power required is only 1% to 2% of that of a similar tunnel using atmospheric air. The oxyfluorides SO_2F_2 and POF_3 and freon 12

Royal Aircraft Establishment, Farnborough, Report No. Aero 2081, August 1945.

(CCl_2F_2) are next in order of economy; they require about 5% of the power with air. All have the very desirable feature of being chemically inactive, almost inert substances. The linear dimensions of tunnels using them are about $\frac{1}{4}$ to $\frac{1}{3}$ of those of the corresponding air tunnels.

The value of γ for all these substances is low -- about 1.15, compared with 1.4 for air. No substance is known which will permit substantial power reduction at normal temperature with $\gamma = 1.4$.

If $\gamma = 1.4$ is essential -- nothing definite can be said on this point -- then power economy is best achieved by refrigeration. This is permissible down to a definite limiting temperature. For air the limit is 126° Absolute, and the power there is only 7% of that at normal temperature. Use of nitrogen permits an operating temperature of 108° absolute, and the power required is 3.8% of that for air at normal temperatures. Hydrogen would reduce the power to 1/2%, but there is danger of explosion.

The ultimate extreme in power economy is obtained by operating helium ($\gamma = 1.66$) at a temperature of 7° Absolute. This would reduce the power to 0.00019 of that of normal air (1 H.P. Instead of 5000 H.P.) but the very low temperature presents many problems.

1. Introductory

The earliest high-speed tunnels were designed with the primary object of reaching Mach numbers near unity, and the Reynolds number of the test was low -- usually about 1 to 2 millions. It is now realised that the phenomena which appear near $M = 1$ are associated with complex interactions between shock waves and boundary layers, in which Reynolds number as well as Mach number plays a part. In flight, the Reynolds number of a typical fighter wing, flying near the speed of sound, is about 30 millions, and a power of about 200,000 H.P. Would be required to reproduce it in a wind tunnel operating with atmospheric pitot pressure. (Operating at lower pressures demands still greater powers to achieve the Reynolds number -- see para. 2.1 below.) In practice the high-speed tunnels in operation in U.S.A. and Germany have compromised usually on a maximum R of about 8 - 10 millions, and a power of about 15,000 H.P. on the average.

These very large powers are a great handicap to high-speed research. The expense and unwieldiness of the resulting equipment are serious difficulties in themselves; they are necessarily limited to only one or two in number, and their operators have to devote a large part of their attention to problems of heavy engineering rather than aerodynamics. There are, however, possible methods of making very great reductions in the power required for such tunnels, by choice of new working substances or extremes of working temperatures. It is the purpose of this report to indicate these possibilities broadly, without entering too deeply into the problems of practical application.

2. General considerations

2.1 Power requirements

The power P required to operate a tunnel of working section area A at speed V is given by

$$P \;=\; \tfrac{1}{2}\rho V^{3} \cdot A \cdot \eta$$

where η is a tunnel "power factor" and is a function of Reynolds number and Mach number when the geometry of the tunnel is fixed.

A wing of area S, chord c in this tunnel would operate at the following values of Reynolds number and Mach number:

$$R \;=\; \frac{\rho V o}{\mu}$$

$$M = \frac{V}{a}$$

where

$$a^2 \;=\; \frac{\gamma p}{\rho} \;=\; \frac{\gamma K T}{W}$$

for a perfect gas.

Here, μ, W, γ are the viscosity, molecular weight and specific heat-ratio of the working substance, and K is a constant.

The power required can be expressed in terms of R, M and the properties of the working substance as follows:

$$\begin{aligned}
P \;&=\; \tfrac{1}{2}(\rho V)^2 \cdot \frac{V}{\rho} \cdot A \cdot \eta \\[4pt]
&=\; \tfrac{1}{2}\left(\frac{\mu R}{c}\right)^2 \cdot a \quad M \cdot \frac{a^2}{\gamma p} A \eta \\[4pt]
&=\; \left(\frac{\eta}{2} \cdot \frac{A}{c^2} \cdot R^2 \cdot M\right)\mu^2 a^3 / \gamma p
\end{aligned}$$

In this expression, the term in brackets is specified by the geometry of the wind tunnel, and by the required Mach number and Reynolds number. The quantity $\frac{A}{c^2}$ is the ratio of tunnel to model dimensions and must be kept constant to maintain constant corrections, although the absolute dimensions of both tunnel and model may vary as the working substance is changed. It will be seen that the relative power requirement with changing conditions of the working fluid is measured by the value of $\mu^2 a^3 / \gamma p$.

2.2 Change of working section pressure p

As the power required is inversely proportional to the pressure, it appears at first sight that the wind-tunnel should be operated at the highest possible pressure. There are two reasons why this procedure is unwise. The first, and by far the most important, is that the problem of supporting the model becomes steadily more difficult with increasing pressure.

The force on the wing, at fixed M and R, is proportional to $\frac{1}{2}\rho V^2 S$. This must be carried by supports of length l and cross-section area S', which for a constant loading (similar material) must satisfy

$$\left(\frac{S'}{S}\right) \text{ proportional to } \tfrac{1}{2}\,\rho V^2, \text{ in tension}$$

$$\left(\frac{S'^2}{l^2 S}\right) \text{ proportional to } \tfrac{1}{2}\,\rho V^2, \text{ in compression.}$$

Here l^2 will be proportional to A and S, for similar geometry; so that

$$\left(\frac{S'}{S}\right)^2 \text{ proportional to } \tfrac{1}{2}\,\rho V^2, \text{ in compression.}$$

Thus in either case, minimum strut size requires a minimum value of ρV^2. This requires minimum $\gamma \cdot p \cdot M^2$, so that at constant M a minimum value in γp is required.

Already, with working section pressures less than an atmosphere, high-speed wind tunnels are finding the problem of supporting the model with small interference a major one; and then it follows that power economy by increase in pressure, necessitating increase in size of supports, is highly undesirable.

A second, less important reason, for avoiding high pressures is that the resulting large loads on the wind-tunnel walls require a very much more costly form of construction, which may offset the economy effected by power reduction.

It has been assumed in the following paragraphs, therefore, that p will be about 1 atmosphere -- the simplest structure is obtained if the pressure is rather less than 1 atmosphere in the working section. The relative merits of different working conditions is then given by the value of $\mu^2 a^3 / \gamma$, a simple expression dependent only on the working fluid and its temperature.

2.3. Choice of working fluid at normal temperature

The speed of sound, a, which enters as a cubic factor in the power criterion, varies in proportion to $\sqrt{\frac{\gamma}{W}}$ for perfect gases at constant temperature. Thus for minimum power a substance with a high molecular weight W should be chosen.

Further, the viscosity, which appears as μ^2 in the criterion, in general falls as the number of atoms in the molecule increases (see para. 3.1 and fig. 2). This again points to the use of a substance with large heavy molecules.

There are two objections to this reasoning. Firstly, the substances with large heavy molecules have generally high boiling points; and as it is essential that the working substance should behave identically with atmospheric air (i.e. as a practically perfect gas), its temperature must be well above its boiling point. This requirement sets an upper limit to the molecular weight which can be used; and the problem of a suitable working substance is presented as one of finding a substance with an unusually low boiling point for its molecular weight. A range of possible substances is considered in para. 3 below.

The second difficulty with large molecules is that the value of γ is low. In general, γ is equal to $\frac{2n + 3}{2n + 1}$ where n is the number of atoms in the molecule, so that to reproduce air conditions accurately ($\gamma = 1.4$) the substance must be diatomic. For some of the most attractive substances given in para. 3, the value of γ lies between 1.1 and 1.2, and a few are monatomic with $\gamma = 1.66$. There is no good evidence on the effect of changing γ; surface waves, for which $\gamma = 2$, behave in a manner essentially similar to that of air, so that it is quite possible that changes in γ have no signicance. In support of this, the von Karman solution of the hodograph equations for compressible flow[1], which in essentials consisted in setting $\gamma = -1$, gave very good agreement with the available experimental results for air. The final decision on the effect of varying γ, however, must depend on the results of experiments; and tunnels using the working substances described below provide an economical means for such experiments.

2.4 Effect of change in working temperature

If the temperature is reduced, both the speed of sound and the viscosity decrease, so that the power required to operate at fixed Reynolds and Mach numbers decreases appreciably. For a perfect gas, the variation of power with temperature can be expressed qualitatively:

speed of sound a varies as $T^{1/2}$

viscosity μ varies as $T^{3/2}/(T + C)$

where T is the absolute temperature and C is Sutherland's constant, which is known for a few simple gases and varies from 56 to nearly 500.

It follows that the power required will decrease with decreasing temperature according to the relation

$$P \text{ proportional to } T^{9/2}/(T+C)^2$$

This relation is shown in Fig. 1 for helium, for which $C = 80$; and for air, with $C = 124$. The very great power reduction which is possible by temperature decrease is shown clearly by this figure.

There is a limit to the possible reduction in temperature; it is essential that the fluid should not reach its boiling point at any part of its circuit around the wind tunnel. The lowest temperatures are attained at the model, at the points of highest velocity. Here it is found experimentally that a Mach number in excess of 1.4 is seldom obtained when the stream velocity is subsonic; and this fact permits the lowest temperature to be calculated with fair accuracy. Writing it as T, and the stream temperature as T_0, and the corresponding Mach numbers as M and M_0:

$$\frac{T}{T_0} = \left(\frac{\gamma-1}{2} M_0^2 + 1\right)/\left(\frac{\gamma-1}{2} M^2 + 1\right)$$

The worst case occurs when a high local Mach number M occurs at a low stream Mach number M_0, as might be found on an aerofoil near its maximum lift coefficient. Thus, for an extreme case, take $M_0 = 0$, $M = \sqrt{2}$. This gives the simple relation

$$T_0 = \gamma \cdot T$$
$$= \gamma \cdot BP$$

i.e., the temperature of the stream in the working section must not be lower than γ times the boiling point, if condensation effects are to be avoided.

This criterion has been used in the following discussion. It is possible that a further criterion based on proximity to the critical point may be desirable. Although there is no fundamental objection to working the fluid in the vapour phase, provided that the thermodynamic properties remain fairly constant over the working range, yet it is obviously more satisfactory to be sufficiently high in temperature to be certain that the fluid is acting as a perfect gas. As many of the substances considered below are not included in available tables of critical constants, however, no attempt has been made to examine this futher criterion.

2.5 Tunnel dimensions -- general relations

Any change in working conditions or fluid will in general change the absolute dimensions of wind-tunnel and model, to achieve the same values of R and M. As there are quite definite limits to the size of tunnel which is practically convenient, some attention must be given to this. Present high-speed wind tunnels have working sections about 9 - 16 ft. diameter; they cannot be increased in diameter without great increase in operating difficulty. On the other hand, if diameter is decreased the precision required in model-making becomes greater until accuracy must necessarily be lost. This probably occurs at about 1 ft. diameter, if not more. Thus the practical range can be regarded as from 1 to 10 ft. in diameter.

The effect of change of working substance can be seen from the equations of para. 2.1. We have

$$c = \frac{R\mu}{\rho V} = \frac{R}{M}\frac{\mu}{\rho a}$$

$$= \frac{R}{M}\cdot\frac{\mu a}{\gamma p}$$

Thus for wind tunnels operating at the same values of R and M, and at the same working pressure, the linear dimensions are proportional to

$$\frac{\mu a}{\gamma}$$

By comparison with the power criterion, the tunnel dimensions will in general decrease as the power is reduced. They will also decrease with decreasing temperature, in proportion to

$$T^2/(T+C).$$

3. Consideration of special working substances

3.1. Derivation of data

For a few simple substances, the values of μ, a, and γ and the boiling point have all been measured. For many substances likely to make useful fluids for tunnel use, however, the necessary data has had to be derived.

The speed of sound, a, can very easily be obtained from the value of γ and the molecular weight; in fact, for all the substances with known values of a, the following relation holds very accurately:

$$a \text{ (in ft./sec.)} = 5030\sqrt{\frac{\gamma}{W}}$$

The value of γ, where it has not been measured, has been taken to be $(2n + 3)/(2n + 1)$ where n is the number of atoms in the molecule. This may be inaccurate in the case of some substances with large n, particularly in the vapour stage.

Greatest doubt applies to the estimation of viscosity μ. It is observable that values of μ, for those substances on which measurements have been made as a gas or vapour, fall steadily as the number of atoms in the molecule increase. This is shown in Fig. 2, in which all available values have been plotted. The mean line drawn in this figure has been used to give a rough estimate of viscosity for other substances.

Boiling point has been measured for all the substances considered. The value of Sutherland's constant C is known for a few simple substances; with more complex substances it becomes very large, and is only required to very small accuracy. A very rough estimate based on known values is then sufficient.

3.2. Results

Table 1 gives the significant physical constants for some substances (estimated values, as distinct from measured, are indicated by the letter E) and the resulting values of the power criterion $\frac{\mu^2 a^3}{\gamma}$ and the "dimension factor" $\frac{\mu a}{\gamma}$. Two cases have been considered; that of a tunnel operating at a normal temperature, viz. $288°$ absolute, in the working section, and that of a refrigerated tunnel operating at the minimum permissible temperature of $\gamma \times$ B.P. A few substances have boiling points which make $288°$ below the permissible value, but are nevertheless quite attractive if operated at a slightly higher temperature; these have been included, but in general substances requiring very high tempeatures to operate satisfactorily have been ignored.

Considering first the wind tunnel operating at normal temperatures (288° Absolute), the table shows several substances will permit the power to be reduced to less than 5% of that required by air at the same temperature. Without exception, these are fluorine compounds; in fact, fluorine compounds in general appear to have unusually low boiling points for their molecular weight. They have the further advantage of being, on the whole, comparatively inactive chemically. Thus the oxy-fluorides, SO_2F_2 and $P\ O\ F_3$, are chemically inert, and are generally comparable with nitrogen in this respect. The freons CCl_2F_2 etc., have been adopted commercially in America as refrigerants, because they contrast very favourably with other refrigerants in their lack of chemical activity. The hexa-fluorides are also described as extremely stable, unreactive substances.

From the point of view of power economy at normal temperature, the hexa-fluorides easily take first place. The compound SF_6 requires only 2% of the power required by air; SeF_6 and $Te\ F_6$ are still better, the latter requiring only 1% of the power of air. As the last two are probably difficult to obtain in quantity, however, sulphur hexa-fluoride appears to be the most promising. Arsenic fluoride, $As\ F_5$, is slightly better, but nothing is known of its chemical properties. All have boiling points sufficiently low (about 220 - 240° Abs.) to be well clear of condensation troubles at 288° Abs.; in fact operation at some 20 - 30° below this temperature is permissible, and the power requirement is then reduced to about 1% for all four fluorides.

The oxy-fluorides SO_2F_2 and $P\ O\ F_3$, and the simplest freon $C\ Cl_2F_2$, all require some 5% of the power required for air. Other freons, $C_2\ Cl_2\ F_4$ and $C_2\ Cl_3\ F_3$, require higher temperatures to avoid condensation effects, and they can then give rather better economy than $C\ Cl_2\ F_2$. In U.S.A., where these freons are available commercially, they have already been proposed as the working substance for a wind tunnel, and have been used in some basic research work[2] at high Mach number and Reynolds number. Where none of these substances is available commercially, the hexa-fluorides would seem to be preferable.

For all these substances, the size of the wind tunnel is reduced to about 20 - 35% of the linear dimensions of a similar tunnel using air. This implies a diameter of about 2 to 4 ft. for a Reynolds number of 10 millions; a quite convenient size. The corresponding power, in the best case, would be about 200 H.P. as compared with 15,000 H.P. for air.

The value of γ, for all these substances, lies between 1.13 and 1.18. There is no comparable diatomic substance, so that it is not possible to achieve similar reductions in power whilst still retaining $\gamma = 1.4$. The best which can be done is to use hydriodic acid (HI) at 332°, which requires a power of 18% of that for air; this is not recommended, however, in view of the highly corrosive property of the acid.

To achieve low power with $\gamma = 1.4$, it is necessary to refrigerate. As several of the high-speed tunnels now in existence already have large refrigerating plants, the proposal to operate at very low temperatures does not mean a radical change; and the power absorbed in refrigeration will still remain, as at present, as a small fraction of that required to operate the tunnel, except in a few extreme cases pointed out below.

If the air is cooled down to 126° Abs., which is sufficiently high to ensure that the oxygen component does not condense near the model, then the power required is reduced to 7% of that at normal temperature. With nitrogen, a rather lower temperature of 108° Abs. is permissible; this results in better economy (power abosrption of 3.8%) and a safer wind tunnel operating on inert gas. The linear dimensions of such a tunnel, one-quarter of those of the corresponding tunnel at normal temperatures, are very convenient; and the value of γ is identical with that for atmospheric air.

At extremely low temperature, hydrogen is the most attractive diatomic substance from the power economy aspect. It can be operated at 28° absolute, and its power requirement is then only 0.0048 of that of air at normal temperature. It is not an ideal substance for a tunnel, however, in view of the danger of explosion.

Carrying temperature reduction to the extreme, helium ($\gamma = 1.66$) is by far the most spectacular substance from the point of view of power economy, as a result of its very low boiling point. It is permissible to operate with helium at temperatures down to 7° absolute, at which point the power, compared with air, is reduced in the ratio 1.9×10^{-4} to 1. A helium wind tunnel giving the same Mach number and Reynolds number as the R.A.E high speed tunnel (4,000 H.P., 10 ft. x 7 ft. working section) at the same pressure, would have a working section of 2.3 inches x 1.6 inches, and would require a maximum power of three-quarters of a horse-power!

A wind tunnel of this type, however, presents many problems. Substances capable of taking the loads on the model have to be found; and the tunnel itself is too small to permit accurate model-making without great cost. The worst problem is that of refrigeration to such a low temperature, and absorption of the operating power. This is helped by the small size of the tunnel and the small power; but even so, as conventional refrigeration methods have to be abandoned in favour of porous plug technique, the problem is a large one. To a lesser extent, these comments also apply to the use of hydrogen as suggested above. The cost of a helium tunnel as indicated here is so small, however, in comparison with its equivalent using atmospheric air, that some consideration of its problems may be worth while.

References

No.	Author	Title, etc.
1	Karman	Compressibility effects in aerodynamics Journal of Aeronautical Sciences, July, 1941.
2	Theodorsen, Regier	Experiments on drag of revolving discs, cylinders, and streamline rods at high speeds. N.A.C.A. Advance Confidential Report, June, 1944.

Attached:

Table 1
Figs. 1 and 2 Drg. Nos. 1757OS and 1757IS

Circulation:

(Part omitted)

Table 1

Properties of some substances suitable for tunnel; power required and dimensions relative to air at 288° Abs.

E indicates estimated value; N indicates operation not possible at 288° A

Substance	Formula	Boiling Point °Abs.	γ	At 288° Absolute				At minimum temperature			
				Speed of sound a ft/sec.	$\mu \cdot 10^{-7}$ $\frac{slug}{ft.sec.}$	Power relative to air $\left(\frac{a^3\mu^2}{\gamma}\ ratio\right)$	Dimension relative to air $\left(\frac{a\mu}{\gamma}\ ratio\right)$	Min. temp. $\gamma \times$ B.P. °Abs.	Sutherland constant C	Power relative to air at 288° Abs.	Dimension relative to air at 288° Abs.
Air	--	90	1.40	1117	3.79	1.00	1.00	126	124	0.069	0.31
Helium	He	4	1.66	3270E	4.12	25.2	2.69	7	80	1.9×10^{-4}	0.019
Neon	Ne	27	1.66	1460E	6.52	5.57	1.90	45	56	0.015	0.16
Krypton	Kr	120	1.66	705E	5.14	0.39	0. 72	199	188	0.15	0.42
Xenon	Xe	165	1.66	560E	4.64	0.16	0.52	274	252	0.13	0.48
Hydrogen	H_2	20	1.40	4240	1.86	13.2	1.86	28	72	0.0048	0.063
Nitrogen	N_2	77	1.40	1130E	3.64	0.96	0.97	108	110	0.038	0.25
Carbon monoxide	CO	82	1.40	1135	3.60	0.95	0.97	115	102	0.049	0.28
Oxygen	O_2	90	1.40	1070	4.14	1.05	1.05	126	127	0.073	0.33
Chlorine	Cl_2	239	1.40	700	2.72	N	N	334	325	0.41	0.56
Hydriodic acid	HI	237	1.40E	529	3.86	N	N	332	390	0.18	0.60
Sulphurated hydrogen	H_2S	214	1.34	1010E	2.59	0.36	0.65	287	--	0.36	0.65
Carbon dioxide	CO_2	195	1.30	868	3.01	0.32	0.67	253	240	0.21	0.55
Nitrous oxide	N_2O	183	1.32	875	3.01	0.32	0.66	242	313	0.18	0.75
Sulphur dioxide	SO_2	262	1.26	707	2.59	N	N	330	--	0.21	0.59
Selenium hydride	H_2Se	231	1.29E	633E	2.80E	N	N	298	--	0.12	0.47
Sulphur oxyfluoride	SOF_2	243	1.22E	598E	2.52E	N	N	296	--	0.085	0.42
Antimony hydride	SbH_3	255	1.22E	503E	2.52E	N	N	311	--	0.061	0.38

Table 1 (continued)

Substance	Formula	Boiling Point °Abs.	γ	At 288° Absolute				At minimum temperature			
				Speed of sound a ft/sec.	$\mu \cdot 10^{-7}$ $\frac{\text{slug}}{\text{ft.sec.}}$	Power relative to air $\left(\frac{a^3\mu^2}{\gamma}\ ratio\right)$	Dimension relative to air $\left(\frac{a\mu}{\gamma}\ ratio\right)$	Min. temp. $\gamma\times$B.P. °Abs.	Sutherland constant C	Power relative to air at 288° Abs.	Dimension relative to air at 288° Abs.
Methane	CH_4	109	1.31	1450	2.21	0.80	0.81	143	--	0.079	0.30
Methyl chloride	CH_3Cl	249	1.18E	775E	2.25E	N	N	294	--	0.15	0.50
Carbon tetrafluoride	$C F_4$	258	1.18E	583E	2.25E	N	N	304	--	0.072	0.40
Sulphur dioxyfluoride	$S0_2F_2$	221	1.18E	543E	2.25E	0.049	0.34	261	--	0.035	0.29
Phosphorus oxyfluoride	POF_3	233	1.18E	538E	2.25E	0.047	0.34	275	--	0.040	0.32
Silicon fluoride	$Si F_4$	208	1.18E	538E	2.25E	0.047	0.34	246	--	0.029	0.27
Chloroform	$CHCl_3$	334	1.15	492E	2.13	N	N	371	454	0.083	0.44
Freon 12	CCl_2F_2	243	1.18	485	2.58	0.047	0.35	280	--	0.042	0.34
Carbon tetrachloride	CCl_4	350	1.13	430E	2.09	N	N	396	--	0.069	0.44
Ethylene	C_2H_4	170	1.26	1070E	2.13	0.31	0.60	214	226	0.133	0.38
Trifluorethylene	C_2HF_3	222	1.15E	593E	2.10E	0.056	0.36	255	--	0.038	0.30
Arsenic fluoride	$As F_5$	220	1.15E	412E	2.10E	0.019	0.25	253	--	0.013	0.21
Sulphur hexafluoride	$S F_6$	222	1.13E	442E	1.96E	0.020	0.26	251	--	0.013	0.21
Selenium hexafluoride	$Se F_6$	234	1.13E	387E	1.96E	0.014	0.22	265	--	0.011	0.19
Tellurium hexafluoride	$Te F_6$	237	1.13E	342E	1.96E	0.010	0.20	268	--	0.008	0.18
Freon 113	$C_2Cl_2F_4$	321	1.12E	407	2.11E	N	N	360	--	0.042	0.36
Freon 114	$C_2Cl_3F_3$	277	1.12E	382	2.11	N	N	310	--	0.020	0.27

Table 1 (continued)

Substance	Formula	Boiling Point °Abs.	γ	At 288° Absolute				At minimum temperature			
				Speed of sound a ft/sec.	$\mu \cdot 10^{-7}$ $\dfrac{slug}{ft.sec.}$	Power relative to air $\left(\dfrac{a^3\mu^2}{\gamma}\ ratio\right)$	Dimension relative to air $\left(\dfrac{a\mu}{\gamma}\ ratio\right)$	Min. temp. $\gamma \times$ B.P. °Abs.	Sutherland constant C	Power relative to air at 288° Abs.	Dimension relative to air at 288° Abs.
Propylene	C_3H_6	223	1.11E	820E	1.74E	0.106	0.43	248	--	0.066	0.34
Methyl ether	C_2H_6O	249	1.11E	780E	1.74E	0.091	0.40	276	--	0.078	0.38
Benzene	C_6H_6	353	1.08	670E	1.55	N	N	382	--	0.099	0.49
Iso-butane	C_4H_{10}	263	1.07E	684E	1.56	0.051	0.33	282	--	0.047	0.32
Ethyl ether	$C_4H_{10}O$	308	1.07	603E	1.50	N	N	330	--	0.052	0.34

REDUCTION IN POWER OF TUNNEL
R, M FIXED
HELIUM
AIR
POWER REQUIRED / POWER AT 288K
1.2
1.0
0.8
0.6
0.4
0.2
0
0
50
100
150
200
250
300
WORKING SECTION TEMPERATURE (K)

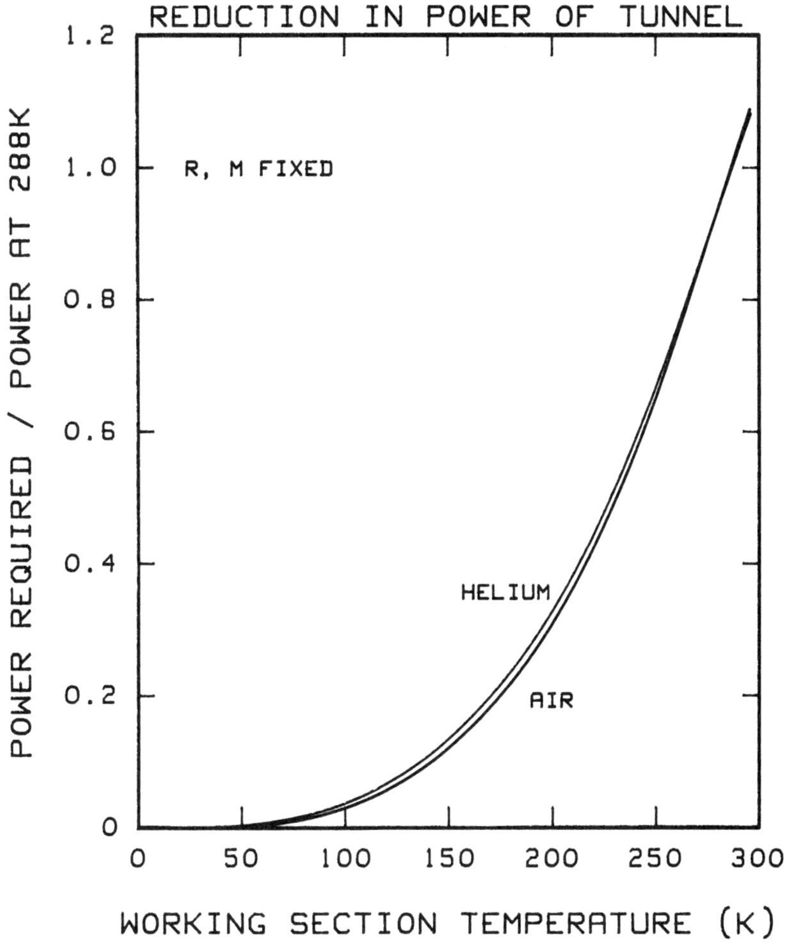

Figure 1

VISCOSITY OF GASES AND VAPOURS
7
6
5
4
3
2
1
0
μ (X10⁻⁷) IN SLUG/FT-SEC UNITS AT 288K
Ne
Kr
Xe
He
O₂
HI
N₂
CO
N₂O
CO₂
Cl₂
H₂S
CCl₂F₂
SO₂
CH₄
CHCl₃
CCL₄
C₂H₄
CCL₃F₃
H₂
C₆H₆
C₄H₁₀
C₄H₁₀O
MEAN CURVE
0   2   4   6   8   10   12   14   16
NUMBER OF ATOMS IN MOLECULE

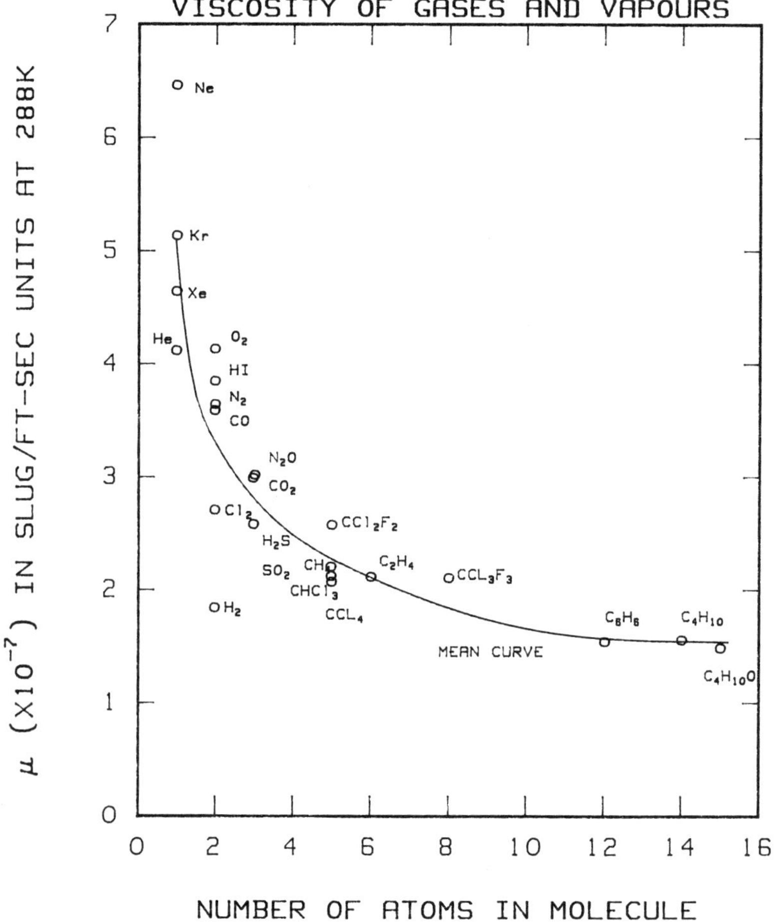

Figure 2